WILLIAM RYAN

AND WALTER PITMAN

ILLUSTRATIONS BY
ANASTASIA SOTIROPOULOS

MAPS BY WILLIAM HAXBY

SIMON & SCHUSTER PAPERBACKS
NEW YORK   LONDON   TORONTO   SYDNEY

Noah's *Flood*

The New Scientific Discoveries

About the Event

That Changed History

SIMON & SCHUSTER PAPERBACKS
Rockefeller Center
1230 Avenue of the Americas
New York, NY 10020

SIMON & SCHUSTER PAPERBACKS and colophon are registered
trademarks of Simon & Schuster, Inc.

For information about special discounts for bulk purchases,
please contact Simon & Schuster Special Sales:
1-800-456-6798 or business@simonandschuster.com.

Designed by Karolina Harris

Manufactured in the United States of America

9   10

The Library of Congress has cataloged the hardcover edition as follows:
Ryan, William B. F.
Noah's flood : the new scientific discoveries about the event that changed history / William
Ryan and Walter Pitman; illustrations by Anastasia Sotiropoulos; maps by William Haxby.
p.   cm.
Includes bibliographical references and index.
1. Deluge. I. Pitman, Walter. II. Title.
BS658.R93   1999
930'.2—dc21   98-45384   CIP

ISBN-13: 978-0-684-81052-2
ISBN-10:   0-684-81052-2
ISBN-13: 978-0-684-85920-0 (Pbk)
ISBN-10:   0-684-85920-3 (Pbk)

The paths going up and down from forests of cedars
all mourn you: the weeping does not end day or night.

The meadows weep; they mourn you like your mother.
The bear, the hyena, the tiger, the deer, the lion,
the wild bull, the ibex—all the animals of the plain cry for you.

The river Ulaj, on whose banks we walked, laments you;
we traveled its banks. The pure stream bewails you
where we filled our waterbag.

G I L G A M E S H , *tablet viii, column i*
(Gardner and Maier, 1985)

Contents

List of Maps

And God said to Noah, "I have determined to make an end to all flesh, for the earth is filled with violence because of them . . . For my part, I am going to bring a flood of waters on the earth, to destroy under heaven all flesh in which there is a breath of life; everything that is on earth shall die . . ."

On that day all the fountains of the great deep burst forth, and the windows of the heavens were opened. . . . The waters swelled so mightily that all the high mountains under heaven were covered . . . for one hundred and fifty days.

GENESIS 6:13, 17; 7:11, 19, 24

Note to the Reader

All dates, including those determined by carbon 14, are given in calendar years.

Witnesses

PADDLING desperately toward land with his whole torso bent in effort, the voyager is astonished to see a forest rising beneath his raft. The canopy of trees, strangely dusted with reddish brown humus, ascends out of the black void into which he and his family have stared with little hope for days on end. He is puzzled. This is not the mirrored reflection of land and sky he has observed on the surface of the lagoon. Nor is it like the lake at quiet twilight when two moons depart the horizon, one climbing, the other falling into the water.

The bow of his craft enters the mysterious forest that punctures the surface of the water. Only once before has he stared down on treetops like this. He had been taken by prospectors up the mighty river flowing from the setting sun into the high mountains in search of some green metallic treasure. The prospectors had brought him to a wild place where foaming white rapids poured steeply down through a narrow and winding gorge, billowing clouds of mist high into the air. He had stood on a terraced recess that sheltered the ruins of an abandoned shrine. He had picked up a smooth boulder, presumably carried up the cliffs from the stream below, and gazed upon a human face in the representation of a fish goddess, oval eyes set far apart and wide lips tightening in a painful grimace, as if possessed of a foreboding. Then looking back in the direction from which he had come, he had seen treetops poking through the spray and had imagined that he was soaring like an eagle above the clouds. Now he again feels himself to be that eagle, trying to land among the trees. Hungry and exhausted, he prays that his resting site will not vanish, forcing him to continue the flight from his drowned home.

The man jettisons his paddle in order to grasp with both hands the tips of the treetops. He forces his way into a dense tangle of stinging branches, intent on looping a lifeline of hemp around the first sturdy trunk of chalky gray bark he encounters. His children jump overboard and thrash into the brush. They pull their emaciated bodies through the thicket until their feet touch firm earth. They scramble from the water, too weakened to cheer. Their tattooed flesh is coated with the debris of the forest carpet: rotting leaves, insects, mushrooms, and the scum of pollen mixed with algae through which they swam.

For nearly a full cycle of the moon the voyagers have been lost and without destination in an unbounded watery cosmos. They had hastily assembled their boat. It carries only the family group accompanied by a few goats. It also transports precious seeds needed to plant new fields, each variety of grain stored in a woven basket folded in hides to keep it dry. They had not thought to bring water since the lake on which they embarked was sweet. Yet during the course of their voyage it mysteriously turned bitter, eventually killing the fish and wild fowl upon whom they depended for sustenance. Corpses stretched in putrid patches as far as the eye could see.

Birds in search of refuge alighted on their boat and the surrounding flotsam: not just the long-legged fowl they had seen pecking at tiny shells along the beach, but birds they remembered from distant hunting trips across open prairie grassland. Swallows hitched a ride in the rigging, departing only when the white peaks of distant mountains appeared on the horizon.

The family gathers on shore with its salvaged cargo. While unloading the raft, they are alarmed to see that the tree to which their vessel is moored is being submerged. Their flight is not over. To escape the unrelenting onslaught of the deluge they must continue their journey on foot. They decide to climb diagonally up the hillside slope, hoping to intersect a river valley that might take them to open meadows where their animals can forage.

The hazelnut in its flattened hairy husk is not yet ripe for picking, nor is the small stony fruit of the hackberry tree. But there are roots, tubers, and grubs for nourishment. On trails smoothed by deer, they spot the telltale scat of bear and panther. The air is fragrant with laurel blossoms. A brook cannot be far away. While they climb in single file, guiding the animals on whom they depend for milk and wool, the deciduous trees thin out and sunlight pours in to warm their bodies.

The trail joins a path that looks as if it has been worn by the travel of other caravans, up a broad valley in the direction of the early morning sun. A sizable river wends its way past them, headed to the lake from which they are fleeing. The band pitches temporary camp to graze the animals. They look back to see if they are being pursued by the water that has covered their entire world.

Their parents had told of their ancestors who had built houses of reed and tilled fields near the edge of the vast lake. Over the span of generations, however, the lake shore had moved away and had become separated from the village by a lagoon and a strip of sand, sculpted into dunes by the wind, where they would go to gather thick grass to weave into mats. The villagers were familiar with the slow rise and fall of the lake as it accompanied the rhythmic changing of the seasons. One could tell the time of year by measuring the height of the ring of living mussels attached to the pilings of the wharf where boats were tied when not out fishing or carrying goods. Occasionally larger craft would arrive from faraway places on the other side of the lake, manned by sailors who spoke only a few recognizable words, those needed for trade.

At first the rise of the water at the lake's edge was barely noticed. No one paid attention until reports arrived from the fishing fleet of a mysterious new sound. It did not take many days for the rumbling to reach the village from the direction of the midday sun. The reverberation rose and fell and rose again. Conditioned as they were to the natural tranquility of the lakeside, to the wash of small waves along the beach, to the rustlings of animals in the brush, and to the stirrings of village life, the new sound was alarming. It penetrated the walls of their huts. It carried the warning of a distant, violent storm whose rains had already begun to expand their lake beyond its shores.

But the rumbling of the storm did not abate, and the downpour never reached the village. Yet its effects were everywhere: the steady submergence of the wharf, the trembling of the ground, and a mounting coastal swell. As time passed, the lagoon washed over the dunes and reattached itself to the lake.

If the storm continued, the village would be directly threatened by the rising tide. A few villagers began to make preparations for retreat. Some loaded dugouts for travel upriver, unnerved by their discovery that the current had mysteriously switched direction and now flowed inland. Others, fearful that the storm might soon beat directly on them and capsize their boats, shunned the water and began to trek across the arid steppe to hunting grounds far from the lake. However, many stayed put. The water had never in memory risen enough to flood their village, built upon the foundations of previous villages and now the highest point of the landscape within many days' walk.

Floods were, in fact, almost always beneficial. They submerged fields planted with seed. If the water did not linger, the next harvest would be bountiful. Yet this storm continued. It grew in intensity. The land slipped rapidly from sight. Those who had not departed at the early warning eventually saw their village transformed into an island that steadily shrank in size. Those commanding large boats for fishing and trade pulled anchor. Panic set in among the remaining population. They began to tear apart houses and sheds to obtain beams and braces to assemble makeshift floats. One by one a number of strange craft and their frightened passengers were carried away by the wave of destruction and desolation. By then a sea of death had consumed all that was familiar.

ELSEWHERE the first warning of the coming disaster was also the whisper of a sound, a weak throbbing from far away. Several bands of farmers on the delta had settled in clusters of post and beam houses surrounded by fields and corrals. They grew grain, raised cattle, fished in the nearby lake, and hunted game in the marsh. Confused by the strange sound, they stood on the bluff and stared in the direction from which it seemed to radiate. They could see nothing except a constant cloud that lit up the night sky.

A group of distant kinsmen, haggard and terrified, came in along the shore. They described how their valley had been inundated. The waters of the Great Salt Sea above were pouring into the lake below. The hollow where they had lived had become a wild flume. Their villages had been swept away by the torrent.

Days later the rising water encroached on the delta and invaded some of the fields nearest the lake through the network of irrigation ditches. The puddle swirled around their dwellings and soon imperiled the granaries. The far-off noise thundered louder and louder while the water mounted every day, bringing with it a wash of floating debris. The frightened villagers headed upriver into the hills. Apprehensive that they were being punished, they made sacrifices to their goddess, praying that this awful nightmare would soon pass. If the waters did not recede, the people pledged to retreat across the mountain divide in search of new land. If necessary they would displace others with whom they had traded and whose shiny black stones they had knapped and then flaked into the sharp blades of their spears.

ON yet another rim of the lake a column of adventurers had made its way toward the noise, forced by the rising water to trek inland from the lagoons where they had once snared waterfowl. They detected a line of flotsam slowly twisting along the shore as it ebbed past them in the opposite direction. The water was muddy. The shore's edge was infested with decaying fish and other carrion. The filthy spume entered the lagoon. The torso of a deer, skin torn, limbs mangled as if mauled by some kind of gigantic malevolent creature, sloshed back and forth in the surf. To avoid the stench, the party mounted the escarpment that overlooked the encroaching coastline. The usually abundant big game animals were nowhere to be seen, perhaps intimidated by the anomalous behavior of the natural world.

Over ground that trembled, the group pressed on toward the sound. The thunder numbed their ears. A chilling salty mist blew in on them. It scattered the sunlight to create an unearthly stage without shadow. From a plateau they gazed in awe at gigantic plumes of spray billowing upward out of the chasm ahead. Out of its hidden bowels belched a hideous roar. At the lip of the precipice the daring few looked down with dismay on a scene of devastation. The landscape of the once fertile valley floor had been obliterated. The walls, which they remembered as steep, wooded slopes covered with foliage, had been stripped naked. Once so peaceful, the valley was now a raw wound filled with a monstrous Leviathan, writhing snakelike toward the lake below.

Where the passage bent sharply, the rushing cascade was colliding with an opposing cliff, eating away the bedrock and leaping upward almost to the plateau on which the wayfarers stood. The surface of the water convulsed in a mass of scaly foam. Whole trees spun in whirlpools, trunks thrown like splinters into the air, then falling out of sight into waves cresting around boulders. A face of the steep slope collapsed in an avalanche that disappeared into the water, completely flushed away.

Far in the distance, where the flume from the ocean met the lake, lay a maelstrom of gigantic twisting eddies of water and wreckage. The jets drove deep into the body of the lake, feeding a huge crater of gigantic whirlpools. Beyond that rose an immense fountain of water from which other waves rebounded. Everywhere above this turmoil, great bursts of spray atomized in a mist that steadily mushroomed into a swirling black cloud illuminated from within by continuous discharges of lightning.

The blast of the torrent beat on the witnesses, overwhelming their senses. Petrified by the joining of the ocean and the lake, they staggered back from the precipice and fled homeward to tell of the chaos unleashed by the gods and the doom it foretold.

I

The Discovery of the Flood Story

CHAPTER ONE

Deciphering the Legend

"We need not try to make history out of legend, but we ought to assume that beneath much that is artificial or incredible there lurks something of fact."

C. Leonard Woolley, 1934

O N an autumn day in 1835, the same month that Charles Darwin finished five weeks of observing finches in the Galapagos Islands, Henry Creswicke Rawlinson made a reckless mistake. He had lain a ladder on its side high on a cliff in the Zagros Mountains of Persia to move laterally along a shelf too precarious to cross by stepping sideways. As Rawlinson walked on the lower side beam and wove his feet between the rungs, his weight tore the brace free. In an instant he and the disintegrating ladder went crashing down the precipice. Weightless, he instinctively reached for the cliff face, the palms of his open hands sliding in vain down the sheer, polished stone. As the British envoy grasped for the slightest crack to arrest his fall, his fingers tightened around something solid. To his astonishment, it was not rock but wood. Miraculously, the two ends of the upper beam had snagged in separate crannies. His fall was arrested.

Only a moment ago Rawlinson had set out to traverse a vertical wall into which were carved column upon column of inscriptions of an ancient language. Above the inscriptions a colossal sculpture adorned this, the world's oldest roadside billboard, the Behistun Rock. Its huge carved panel displayed a mighty king standing in judgment over nine chained captives— the neck of a tenth victim crushed underfoot. The conqueror, with left hand resting on the tip of his bow, held his right arm high in reverence to a

Henry Creswicke Rawlinson falling down the face of the Behistun Rock

winged deity. The oppressive shadows cast down the bas-relief by the noon sun infused the scene with despair.

Rawlinson hastily hauled himself hand by hand back to the security of the arms of terrified friends, who were watching the unfolding drama. He regained his composure, relieved still to be alive and hardly scratched. For months he had been using ladders on this cliff without incident. By now he had copied almost a third of the inscriptions, covering a breadth of sixty feet across the wall. His explorations on this mountain side summoned all his talent as an experienced climber.

Rawlinson had arrived in Kermanshah (today a major city in western Iran) the previous March, in the capacity of a military advisor to the brother of the Shah. He was an officer in the service of the East India Company, to which he had volunteered as a cadet at age seventeen. Born in Oxfordshire, England, in 1810, he spent his early youth at boarding school in Somerset and later earned a scholarship to Ealing School on the outskirts of London. Taking naturally to linguistics, he immersed himself in Latin and Greek, and devoured the works of Homer, Herodotus, Plato, Virgil, and Pliny.

While the adolescent Rawlinson was cutting his teeth on the classics, an event took place on the campus of Oxford University that conditioned the intellectual atmosphere of the natural sciences for the next half century. Over the years it had become widely known that vast tracts of the English and Scottish bedrock were littered with chaotic accumulations of sand, large boulders, and boulder clay. The extreme disorder and unstratified nature of this so-called drift suggested a sudden deposition from powerful currents in rapid movement.

In 1820 the Reverend William Buckland, the most influential geologist in England, gave his inaugural lecture as professor of mineralogy and geology at Oxford. Buckland's speech, entitled "The Connection of Geology with Religion Explained," attributed all of these widely scattered unstratified deposits to the ravages of the universal flood of Noah. Buckland believed fervently that the purpose of his science should be nothing less than

> to conform the evidences of natural religion; and to show that the facts developed by it are consistent with the accounts of the creation and deluge recorded in the Mosaic writings.

Many of the leading geologists in England during the 1820s and 30s, such as Charles Lyell, were honored to be Buckland's pupils. Most people who went to the highlands to observe the drift deposits had only to look with their own eyes to see unequivocal evidence of the violent convulsion of water capable of carrying the gravel and boulders from afar and sweeping marine shells into superficial deposits hundreds of feet above the sea. It was therefore typical at the time of Rawlinson's graduation from Ealing and his departure for service in the East India Company for the naturalist and biblical archaeologist to go out into the world, look for, and find the actions of ancient man and God as chronicled in the Old Testament.

During his four-month voyage by clipper ship to India, Rawlinson became acquainted with the governor of Bombay, who in the course of stimulating dinner conversations inspired in him a passion for the history of Persia, its ancient languages, religions, and sacred literature. Rawlinson heard firsthand of the Indo-European enigma that had been proposed forty years earlier during a celebrated discourse of an English jurist serving in Calcutta. The High Court judge, Sir William Jones, had recounted to the Asiatic Society of Bengal his three-year immersion in the study of ancient legal and religious Hindu texts written in archaic Sanskrit. The oldest of the stories—the Rig-Veda—recounted the heroic traditions of a people who called themselves Aryans and possessed an epic legend of a great flood. In this myth the survivor, Manu, is warned by a fish that "a deluge will engulf all the creatures" and is then instructed to "prepare a ship" that will carry him to "the northern mountain." In the course of his readings Jones recognized a remarkable linkage of common words and grammatical constructions among Sanskrit, Latin, Greek, Welsh, early German, and Persian. The overlap suggested that the separate languages had sprung from some common source that perhaps no longer existed.

Rawlinson quickly realized the implications of Jones's observations. A mother language had existed in the prehistoric past, spoken by unknown people in an unknown homeland who, at some indeterminate date, had dispersed into Europe, Asia, and India. Sanskrit and Persian were daughter tongues of this protolanguage. Enticed by the notion that he might discover the language of Noah's son Japhet and his offspring, who migrated out of Armenia where the ark was thought to have landed, and captured by the prospect of one day unraveling the path by which languages spread, Rawlinson set out to become fluent in Persian, Arabic, and Hindustani as interpreter and paymaster to the First Grenadier Regiment stationed in Bombay.

On an assignment as military advisor to the Shah of Persia with a post at Kermanshah, the inquisitive envoy took the opportunity to visit the imposing ruins of Persepolis, royal city of the Achaemenid Dynasty of Persia, whose rulers' exploits had been so richly narrated by the early Greek historians Herodotus and Xenophon. Here Rawlinson caught his initial sight of cuneiform script, a name derived from the Latin *cuneus* for "wedge" or "angle." A few lines of this undeciphered writing had been impressed onto paper and sent to Europe seventy years earlier.

Some miles to the west on a broad fluvial terrace at the confluence of the Kur and Polvar rivers, Rawlinson wandered in fascination through ruined palaces of fine-grained marble and tall columns, still erect but scarred by fire and pillage. He gazed on the megalithic tombs of the Achaemenian kings.

To Rawlinson, not yet twenty-five, everything was virgin and wondrous. He could imagine that one of the tombs must be the resting place of King Darius I the Great, who seized his throne in the late sixth century B.C. after a violent struggle to contain a palace revolution. Darius governed skillfully and managed a vast empire long before that of Alexander the Great, a regime that encompassed all the prior realms of the Egyptians, Chaldeans, Ionians, Persians, and Medes, extending to the east as far as the Indus Valley, to the west into Europe, and to the south into Africa, flourishing in economy and culture.

Not long after arriving in Kermanshah, Rawlinson enticed a local Kurdish boy to accompany him on horseback. The two rode off on Rawlinson's Arabian stallion for more than twenty miles into the Alvend foothills, whereupon the boy guided his host to the imposing Behistun Rock, which up to then had never been known to the western world. The winding pass they entered had been a route for caravans and armies ever since the dawn of civilization. The immense volume of the inscribed texts on its nearly vertical wall and the question why they had been placed there intrigued Rawlinson. He felt a surge of confidence that these strange signs were indeed a proclamation, broadcasting to posterity the victor's name, his feat of arms, and the genealogy of his family. Rawlinson's intimate knowledge of Greek, Persian, and Hindustani lore would help him penetrate the millennia of ghostly silence.

In the course of the next two years, Rawlinson returned again and again to the Behistun Rock, each time picking up where he had left off in the task of copying the inscriptions. Although he had quickly noted that they were trilingual, he recognized only one of the styles of writing, which he had seen earlier in the ruins of Persepolis. The other two (Babylonian and Elamite) were completely unknown. Focusing at first on the familiar script, inscribed in what he believed to be Zend and possibly the oldest language of Persia, Rawlinson saw a similarity and repetition in certain groups of signs. He remembered a passage from Herodotus where mighty Xerxes exclaims:

> May I be no son of Darius, son of Hystaspes, son of Arsames, son of Ariaramnes, son of Teïspes, son of Cyrus, son of Cambyses, son of Teïspes, son of Achaemenes, if I do not punish the Athenians.

Not only did Rawlinson know these names of the Persian kings, but he was also familiar with their pronunciation in Sanskrit, often with sounds and emphasis different from the Greek that was spoken in classical time. One by one he tried fitting the kings' names with Persian inflection into the inscriptions next to and following the commonly repeated group of signs, all the time searching for a correct match in phonetic value of identical glyphs. With painstaking effort in copying and hours and hours with his notebook at his desk at the British Residency in Kermanshah in 1836, he pried open the immortal inscription whose message had long ago slipped into oblivion:

King Darius proclaims: I Darius, the great King, King of kings, King in Persia, King of countries, son of Hystapes, grandson of Arsames, an Achaemenian.

Rawlinson mailed his decipherment to the Royal Asiatic Society in London, an accomplishment that earned him the recognition of being the first to crack the cuneiform code. Within a few more months he had translated the bulk of the two hundred lines of Old Persian copied from the Behistun Rock. King Darius's proclamations had indeed been honored for twenty-five centuries.

Ye who in future pass, will see this inscription, which I have carved in the rock, of human figures there. Efface and destroy nothing! As long as posterity endures preserve them intact!

In confirming the real places and people of Herodotus's history, Rawlinson added affirmation to a narrative that according to the ancient Greek historian Thucydides had "fought its way into the country of myth." For in the telling of the exploits of King Darius, Herodotus, like the epic poet Homer, had relied almost entirely on oral tradition. If there were written records, he and his contemporaries lacked the language skills to understand them.

From the very moment he deciphered the Old Persian cuneiform code, Rawlinson became keenly aware of his potential to interpret even more ancient inscriptions in parent languages. These accounts, yet to be unearthed from beneath desert sands, could lead him to the authentication of lost cities and to the exploits of their legendary kings—the knowledge of which existed only in Old Testament references. Thirty-seven years later the promise that Rawlinson saw in prying voices from stone was fulfilled by a protégé of his named George Smith.

UNDER the gleam of an oil lamp that barely penetrated the fog seeping through the night into the cramped, unheated, and windowless alcove above the office of the secretary of the Royal Asiatic Society, Smith sat nearly motionless in tireless concentration while examining the fragments of rocks that he had arranged on the table in front of him. The illumination was ideal to cast a diffuse shadow along each indentation of the wedge-shaped imprints scratched into their surfaces thousands of years earlier.

As an apprentice to Rawlinson in the British Museum in London, Smith had taught himself how to reconstitute the medium on which long-deceased scribes in Mesopotamia had preserved their records. Before him lay pieces of their work that had been excavated from a royal library in Nineveh. The inscription under scrutiny had been etched by a reed stylus onto an original rectangular slab of moist mud, somewhat smaller than a book, which had then been sun-dried. Unfortunately, the tablet had subsequently shattered, either from the heat of the conflagration which ravaged the palace of this last great capital of the Assyrian Empire or perhaps during the hasty archaeological excavation of the library when Smith was just a child. The job at hand was to piece together, out of a huge inventory of broken material, the original whole, which when joined would reveal the unabridged content of the ancient message.

Smith was reassembling a huge jigsaw puzzle, taking advantage not only of shape to match individual edges but also the context of words or partial phrases on their surfaces to see if they were related. Just as one might initially sort the pieces of a picture puzzle into piles based on texture, color, or the presence of a straight edge, Smith had previously sorted thousands of fragments in the museum's collection into groups based on a mere partial glimpse of their subject matter. Those thought to record commercial transactions were gathered in one place, those on marriages and other legal contracts in another, and so on. On this damp fall morning in 1872 he was perusing the collection that he was most eager to translate and that he had labeled the *mythological and mythical.*

From the moment he could read, Smith had been obsessed by scripture. In Bible school he learned by rote the narrative stories of the patriarchs Abraham, Isaac, Jacob, and Joseph. His adolescence coincided with the headline-making discoveries of lost cities of the scripture, notably Nimrud, the biblical Calah, and Nineveh, the last and greatest capital of the Assyrian Empire with its extraordinary sculptures depicting the siege of Lachish in Judah, described in the Old Testament verses of 2 Kings. It had never occurred to this precocious child that he might one day challenge the teachings of the Church of England regarding the divine authorship of the Bible.

Smith had a natural talent for deciphering languages. The one on these tablets from ancient Nineveh was Akkadian, a Semitic tongue. Smith took advantage of his familiarity with Hebrew to parse the predominantly phonetic signs of Akkadian into words that had not been uttered in full sentences for more than twenty-five centuries. He was convinced that when the tablets spoke, they would confirm the biblical version of the earth's beginnings.

In the flickering light that morning, the eager apprentice had recognized an enticing shape. Its particular configuration led him to try to insert it between two other fragments that he had laid out on the table among half a dozen others the previous day. Unfortunately, the broken edges of the tablets were no longer as sharp as they had once been. His trial and error approach therefore demanded that he examine the continuity of any rows of the script that had been aligned by his attempted join. His own words reveal what followed next:

> On looking down the third column, my eye caught the statement that the ship rested on the mountains of Nizir, followed by the account of sending forth the dove, and its finding no resting place and returning.

As Smith's astonishment mounted, he shouted for Rawlinson to hurry in and share the excitement. Deciphering the tiny cuneiform signs no bigger than the letters on this page, Smith knew that he had stumbled on a tablet engraved with the familiar tale of the Great Flood. One version of the day's events recounted that in a state of hallucination, Smith dashed into the hallway and in front of the other startled workers on the museum staff began to tear off his clothes.

Apparently what had so deeply moved Smith was the realization that the fragments he had assembled contained an independent version of the biblical deluge. The heathen words told almost exactly the same story as the Hebrew narrative, right down to the selection of a survivor of the deluge through the intervention of a god, the forewarning that gave time to build a wooden ark, the refuge in it of every kind of animal, bird, and reptile, the grounding of the boat on the side of a mountain, the details of dispatching a swallow, raven, and dove to find land, the offering of a sacrifice, and the pledge that the gods would never again return the world to its primeval watery chaos.

In the weeks ahead Smith added a few more pieces of the tablet, all the while pondering how the flood story in the Bible, a work of supposed divine authorship, could be so similar to another one, even older in its redaction and recorded in an alien myth. Might this mean that the Israelites borrowed a narrative they may have heard when held in captivity in Babylon? Or could it be that the real flood had been such a momentous event in prehistory that its remembrance had been independently preserved in the oral tradition of many different cultures?

When satisfied that he had searched through the museum storehouse with sufficient diligence to have found all the available pieces to his tablet, Smith rendered as full a translation as possible. Three weeks before Christmas he presented his monumental finding before "a large and distinguished company assembled in the rooms of the Society of Biblical Archaeology." A public stir had been aroused with the preannouncement in the London newspapers. Those in attendance included William Gladstone, the prime minister of the British Empire, and the dean of Westminster Abbey. The crowd became transfixed as Smith described his discovery.

On comparing his text, copied during the sixth or seventh century B.C. from an earlier source, with other inscriptions from the much earlier era of Sargon I, founder of the first Akkadian Empire around twenty-three hundred B.C., Smith declared that the original composition could not have been more recent than the seventeenth century B.C. and might be much older. Smith then gave a sketch of his translation, whose substance the audience recognized instantly except for a cast of characters with unfamiliar names, a storm of only six days' duration, and gods so frightened by the tempest that they cowered like dogs into submission and sought refuge by ascending to the heaven of Anu. When the tempest abated, these same gods, now ravished by hunger, swarmed like flies over the sacrifice made by the survivor chosen to escape the waters of the flood. After the feast one of the goddesses flung her jeweled necklace into the sky to be the sign of a covenant never again to drown the world.

Smith's listeners were astonished by this direct literary parallel to the rainbow placed in the sky by the Hebrew God Yahweh as his acknowledgment of the sacredness of life. They then learned that the Akkadian myth even expanded on the biblical version. It referred to a famine and a pestilence that preceded the flood as unsuccessful methods for destroying mankind. Smith's translation further described the opening of a great gate, a thunderous noise, and the bestowing of eternal life on the survivor and his wife after returning to their country, a remote place at the source of the rivers.

So as not to upstage the reporting of young Smith's lecture in the newspapers, Rawlinson politely delayed for twenty-four hours giving his own impressions of the implications of the discovery. When interviewed, he emphasized the great commonality of the flood myth just translated by his protégé and a version handed down by the Babylonian priest Berossus in the third century B.C. He then cited evidence that the flood story related by

Smith had referred to the conquest of Babylonia by the Medes, and therefore it may have originally been authored in the third millennium B.C.

The Times article the next morning quoted Rawlinson as estimating 5150 B.C. for the first of the Mesopotamian kingdoms and pointing out that the flood would, of course, have been older. In his opinion "it could hardly be doubted that the account in Genesis was a version of the same legend which had been carried away by the Abrahamic colony in their original migration from Ur of the Chaldees to Harran and Palestine."

CHAPTER TWO

Conversions

O N a late July afternoon in 1837, Louis Agassiz, barely thirty years of age, addressed the annual meeting of the Swiss Society of Natural Sciences, of which he was president. Standing in front of those assembled in the small Alpine town of Neuchâtel, Switzerland, he chose quite impulsively to abandon his prepared lecture on Brazilian fossil fish, as announced in the printed program. Instead he publicly confessed an almost overnight conversion to a new theory that a vast "ocean of ice" had once covered the whole surface of Europe and all of northern Asia as far as the Caspian Sea.

To support his audacious position, Agassiz described excursions by foot into the Alps the previous summer with his friend Jean de Charpentier, many years his senior. Over a period of several weeks, de Charpentier had demonstrated the role of mountain glaciers in grinding and polishing cliffs of sheer granite and in transporting rock debris plucked from these cliffs to faraway mounds, called moraines, sitting at the terminus of the creeping ice.

The astute de Charpentier had noted similar mounds encasing huge boulders at distances reaching fifty miles from the present glaciers. The participants of the Neuchâtel meeting just had to walk outside to see one of the largest known Alpine boulders, the scratched and faceted *pierre à bot,* that would later be described in support of ancient glaciers in Scotland as "a goodly mansion in size" resting on "the polished, striated, and furrowed surfaces" of the bedrock of the Jura Mountains. What other agent but thick flowing ice could have transported and then abandoned the pulverized rock debris that one persistently encountered far up on the walls of now-forested Alpine valleys?

Although Agassiz dropped his theory of a prior glacial epoch like a bomb-

Louis Agassiz descends fearlessly into an Alpine glacier

shell, it fell on closed ears. The intellectual climate in the beginning of the nineteenth century was such that few naturalists anywhere in Europe considered land ice as more than a curiosity of high-altitude mountain climbing. Certainly the ice played no geological role in shaping the landscape and in leaving significant deposits. In the first edition of his *Principles of Geology*, published between 1830 and 1833, Charles Lyell, regarded by historians as the father of geology, had explained the huge "erratic" blocks scattered across the landscape in unstratified clay as stones released from melting icebergs adrift in the Great Flood.

Despite Agassiz and de Charpentier's having built a trusting camaraderie over the years while collecting fossils in the wilderness of the Alps, Agassiz hesitated in believing that glaciers could travel long distances with rocks in their bellies. He insisted on descending with a rope sling deep into the open crevasses of the ice fields. There with his own eyes he could inspect how the soft snow crystallized after burial into a solid mass of finely layered granular ice. The cross-section of the glacier, captured in the sheer blue-banded walls of these deep clefts, showed hundreds of years of snow accumulation, resulting in a body of frozen water that behaved like extremely viscous wax in its descent downhill under the influence of gravity—constantly deforming itself into bends and cracks while cascading at a pace of two hundred feet per year, denuding and harvesting the rock over which it flowed.

Agassiz almost perished from hypothermia in one of these daring excursions when de Charpentier unknowingly lowered his friend into a stream of meltwater draining into a tunnel at the base of the ice flow. The frigid dousing had a profound effect. It showed vividly that the ice was lubricated at its contact with the bedrock. In this way a mammoth glacier could travel the enormous spans required to deliver the so-called drift far away from the mountain chains. Agassiz eventually confessed his conversion to his mentor and, like most converters, became a devout pupil.

But Agassiz was ill prepared for the hostile reception to his announcement at Neuchâtel. Even an impromptu field trip by horse-drawn carriage two days later through the Jura meadowlands to allow his elder critics to inspect the glacially carried boulders scattered among the wildflowers failed to quell the rancor. One participant to the "fiery discourse about a sheet of ice" remembered:

> In general, I was convinced by my short acquaintance with the leading scientists of the party that a great amount of jealousy and egotism existed between them.

The members of the society clung to the biblical deluge origin for the drift deposits, rejecting the arguments of the young naturalist. Nevertheless, Agassiz's *Studies on Glaciers*, published three years later, would reach a much larger readership ready to engage the stimulating new idea of widespread glaciation. The most important member of the opposing establish-

ment was the Reverend William Buckland at Oxford, the fervent cata-
strophist and perhaps the most talked about scientist in Britain.

Buckland's recent magnus opus, attesting to the action of a universal
flood, had been met with critical praise. Although the chief architect of the
catastrophic synthesis and Agassiz had met before and had even deliberated
the ice age theory during an outing in the Alps, a reconciliation of their
enormous differences in the interpretation of the apparent evidence had
been inconceivable until the Reverend was secure on his home turf in the
British Isles. There in the autumn of 1840, following the annual meeting of
the British Association for the Advancement of Science, the Oxford don, in
characteristic fashion with top hat and academic robe, escorted the Swiss
naturalist to a classic site of the "drift" deposits on Blackford Hill, south of
Edinburgh, Scotland. Pointing authoritatively to the unstratified boulder clay
(the inferred deposits of Noah's torrential flood) at his feet, Buckland pronounced that the rocks within contained no scratches from glacial gouging.

However, Agassiz was wary of ambush by Buckland, about whom it was
said that his elegance rolled like the deluge retiring. Instead, Agassiz ushered
Buckland aside to a nearby cliff that neither had visited before and climbed
up to the underside of an overhanging ledge where the rock leaned forward,
forming a sort of vault. There Agassiz brushed the dust from the face of the
stone, exposed a stunning outcrop of striations (the parallel grooves indicating the action of a moving glacier grinding away at the landscape), and
instantly pronounced, "That is the work of ice!" Buckland's conversion from
diluvialism was instantaneous.

THE test of a good scientific theory is its power of prediction. Agassiz had
been able to walk into a virgin locale and know exactly what clue was likely
to demonstrate the veracity of his ice age and where he could find it. Two
decades would expire, however, before Agassiz's new theory of a global ice
age would be widely acclaimed. The delay was to be expected. It would be
another dozen years before a scientific expedition confirmed that the snow
cover of Greenland was indeed only the thin skin of a monstrous ice sheet
that buried practically the entire landmass, and nearly half a century more
before the volume of the Antarctic ice cap could begin to be comprehended.

The most direct casualty of the new ice age paradigm was any further
necessity of expounding a universal flood to elucidate the origin of the drift
deposits. Even the "shelly drifts" along the coastlines of the Baltic could now
be seen as the product of icebergs plowing aground along the rim of a
shallow inland sea. Consequently, reference to the Old Testament flood as a

global phenomenon rapidly vanished from geology lectures and manuscripts. Diluvial theory melted away into oblivion as had the ice from Agassiz's vast "ocean."

But while the geologists abandoned the flood, biblical archaeologists continued to embrace it and even gave it new life. Agassiz's conversion of Buckland took place at the same time that Henry Creswicke Rawlinson was expanding his knowledge of ancient languages. The continuing excavations of ancient cities in the Near East would resurrect the flood as a defining event in the cradle of civilization.

CHAPTER THREE

Visions of Palaces

THE very year of Buckland's rejection of Old Testament diluvialism, the levees along the Tigris River collapsed from a torrential flood. The interior of the city of Baghdad lay beneath a blanket of mud that smothered the streets to the height of a donkey cart. Houses and walls crumbled from the pressure of a slurry that coursed through the old quarter for weeks. Mud-brick buildings floated free from their foundations and then dissolved in the stream of snowmelt from high peaks of the Taurus and Zagros mountains.

Plague ensued in the wake of the watery onslaught. Those who survived the plague faced imminent starvation. The natural calamity whittled a population of one hundred thousand inhabitants down to a few hundred wretched souls, huddled in squalor.

At about this time a young Englishman, Austen Henry Layard, set off on an overland journey to this scourge-infested outpost, the formerly majestic City of the Caliphs ruled by the direct descendants of Ali. The adventuresome Layard would open the eyes and ears of the Victorian world to the lost treasures of the Tigris and Euphrates, the very rivers that, according to Genesis, had flowed through the Garden of Eden. His quest would bring new interest in the biblical flood even as the geologists were casting it aside.

Following in the footsteps of Henry Creswicke Rawlinson, who had preceded him by a dozen years, Layard, too, hoped to reach India. The magnet that drew him was *The Arabian Nights*, a favorite boyhood book. After reaching Jerusalem on the first leg of his trip, against dire warnings from the local British consul and in the company of a single Arab guide, the twenty-two-year-old Layard trekked self-confidently through the domain of Bedouin cutthroats and highway thieves. On Layard's horizon stretched the

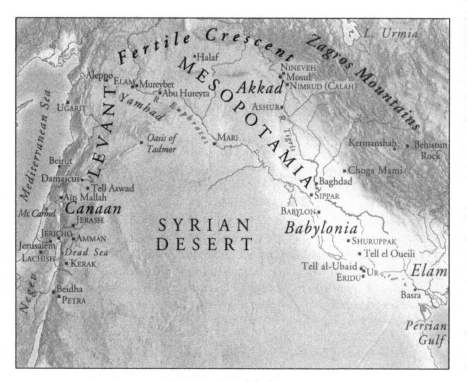

Mesopotamia and the Levant

dust-covered ruins of ancient biblical cities: Petra, Kerak, Jerash, Amman, and countless others. For safety he adopted a strategy to travel unencumbered and without valuables.

> I look back with feelings of grateful delight to those happy days when, free and unheeded, we left at dawn the humble cottage or cheerful tent, and lingering as we listed, unconscious of distance and of the hour, found ourselves, as sun went down, under some hoary ruin tenanted by the wandering Arab, or in some crumbling village still bearing a well-known name.

Though often preyed upon, beaten, and robbed, Layard managed to persuade local caliphs to return at least his notebooks if not his tattered clothes. In Mosul, a walled city of the early Christian era located in present-day Iraq, the British consul, himself a budding Assyriologist, guided Layard to the tell of Ashur. En route, the party pitched its tents beside a mound of substantial height. In the late afternoon, after climbing to its summit and peering across the waterway of the Tigris, Layard was able to discern on the faraway horizon the profile of an even larger edifice. But little did he suspect that this somber silhouette was the city of Calah featured in Genesis and a vault of the most astounding riches from the days of Assyrian splendor. Layard composed his first impressions in his journal.

> The scene around is worthy of the ruin he is contemplating; desolation meets desolation; a feeling of awe succeeds to wonder; for there is nothing to relieve the mind, to lead to hope, or to tell what has gone by.

After a brief respite in Baghdad, which he described as an assemblage of mean mud-brick buildings and a heap of ruins, the adventuresome Layard hitched up with a caravan heading east across the Zagros Mountains into Persia, to Kermanshah. Rawlinson's freshly translated inscriptions of King Darius's proclamation on the Behistun Rock and the still-undeciphered Babylonian and Elamite inscriptions so captivated the intrepid sojourner that he abandoned his goal of India to search for Assyrian antiquity.

> My curiosity had been greatly excited, and from that time I formed the design of thoroughly examining, whenever it might be in my power, these singular ruins.

A year passed before Layard returned to Mosul, where he found a bustle of interest in the ruins of the ancient tells, especially by rival French archaeologists. Not to arouse suspicion and apprehensive that his lack of professional credentials might deny him opportunities for his own digging, he hired a well-known sportsman to assemble munitions, retrievers, and staples for an alleged hunting trip. Without drawing suspicion, he cautiously slipped a few shovels into his wagon and selected a stonemason for a driver. Instead of heading into the foothills, he diverted to an obscure track that followed the meandering Tigris downstream toward the tall, dark mound etched in his memory. His state of anxiety reached such a frenzy that he could not sleep.

> Visions of palaces underground, of gigantic monsters, of sculptured figures, and endless inscriptions floated before me.

The towering tell no longer occupied a riverfront vantage. During the span of its decay, the Tigris River had broken out of the confines of its levees and had cut a new channel almost a mile to the west. As soon as the digging

Austen Henry Layard dusts the debris of the millennia from a huge bas-relief image within the palace of King Assurbanipal at Nimrud

started, however, the old streambed was exhumed, its former left bank being the actual ramparts of the city wall. Layard could see that the main buildings of the city center occupied a platform one thousand feet in width and twice that in length. To the north lay the crumbling Ziggurat, a huge pyramid-shaped temple. Layard's initial trenches attacked two jagged protrusions, parched and wasting in the dry heat of autumn.

The contingent of laborers soon grew to thirty Bedouin Arabs. In the second week, the cornices of two massive slabs emerged from the chamber being excavated. As the slabs were freed from entombment inch by inch, a band of cuneiform inscription became discernible. Then a suite of bas-relief sculptures emerged. In the span of a few hours, trowel and brush opened a panoramic view of battle scenes encircling this very site, the second ruling capital of Assyria. As the dust was swept from the immense stone panels of this city called Nimrud, the name given to the great-grandson of Noah, Layard found himself enraptured by sieges of fortified enemy cities, with row upon row of the defeated being led into captivity, the infantry fording swollen rivers on inflated goat skins amid swimming horses and leaping fish, archers aboard chariots dashing across the open plains in the pursuit of hunted prey, and a bizarre anthropomorphic rendition of an eagle-headed deity over seven feet high reminiscent of the god Nisroch in Genesis.

In the weeks that followed, shrouds dropped from stalking lions, winged bulls, and larger-than-life-size human figures. These scenes were all hastily sketched by Layard and then buried again beneath the rubble. Layard, aware of the significance of his findings, decided he must set out for Baghdad to rendezvous with Rawlinson and enlist his help in protecting the treasure from grasping local officials and rival teams of archaeologists. The assurance Layard sought came in the form of a *firman,* or permit, from the governor.

The next spring a revitalized Layard moved his camp from the filthy mud-brick huts of the local village out onto the yellow-green floodplain budding with wildflowers. He jotted in his notebook:

> As the sun went down behind the low hills which separate the river from the desert—even their rocky sides had struggled to emulate the verdant clothing of the plain. . . . The great mound threw its dark shadow far across the plain, now glittering with innumerable fires. As the night advanced, they vanished one by one until the landscape was wrapped in darkness and in silence, only disturbed by the barking of the Arab dog.

At nightfall lions would often claim command of the ruins, from which, singly or in pairs, they hunted the village dogs. Their success would be heralded by a loud yelp, punctuated by a Bedouin curse. One day the diggers unearthed a human head atop a colossal winged bull. Layard was summoned to the scene. Arriving to a frenzied commotion, the expedition leader brought calm as he studied the stately stare of the bearded creature. More soil was removed from the mammoth torso. A truly frightful creature began to emerge. Could it be a guardian of the main gateway?

His imagination racing, Layard paced off three large strides away from the wall in which the bull was carved. He then scratched the ground with a hoe and instructed the diggers to start afresh at the new spot. The twin of the colossus appeared. Awestruck, the workmen and a growing crowd of onlookers fled from the site in terror.

Although successful in his deduction of a main entranceway, Layard began to feel inadequate and frustrated by his inability to comprehend the fuller history of the city he was unearthing. He was also irritated by the cool reception of the British Museum to his repeated requests for help with his excavation, particularly with the need for professional sketching and on-site preservation. In times of insecurity and depression he sought the encouragement of Rawlinson in Baghdad, who was now preparing to copy the second language inscribed in the Behistun Rock. This writing was thought to be Babylonian, the dialect spoken throughout Mesopotamia long before the Old Persian of King Darius's era. Rawlinson suggested to Layard the possibility that the cuneiform inscriptions in the Nimrud palaces might well serve as a record of the time when the city had been constructed.

When Layard roamed through the twenty-eight royal halls and chambers of his unearthed Nimrud, some with fresco-adorned plaster walls, others with row upon row of sculptured slabs, all deep in the ground, he still did not know the identity of the Assyrian king whose capital this was. Yet the palace floors, paved with burnt brick, were stamped throughout with his very name, which Layard could not decipher. Had Layard only been able to read the accompanying inscriptions, he would have learned firsthand the brutal sadism of Ashurnasirpal II (883–859 B.C.):

> I built a pillar over against his city gate and I flayed all the chiefs who had revolted, and I covered the pillar with their skin. Some I walled up within the pillar, some I impaled upon the pillar on stakes, and others I bound to the stakes about the pillar. . . . And I cut the limbs off the officers, of the royal officers who had rebelled. . . .
>
> Many captives from among them I burned with fire, and I took many as living captives. From some I cut off their noses, their ears, and their fingers, of many I put out their eyes. I made one pillar of the living and another of their heads, and I bound their heads to tree trunks around the city. Their young men and maidens I burned in the fire.
>
> Twenty men I captured alive, and I immured them in the wall of his palace. . . . The rest of their warriors I consumed with thirst in the desert of the Euphrates.

These days of digging were perhaps the most productive in the history of Assyriology. Among the antiquities Layard discovered was the Black Obelisk of the cruel king's successor Shalmaneser III (858–824) (who is mentioned in 2 Kings 7 and 8 of the Old Testament). The intricate bas-relief of all four of its facets portrayed foreign dignitaries delivering tribute to an Assyrian overlord. Alien to Layard at the time, the lengthy text beneath the supplicant figures spoke of the delivery of gold and silver from Jehu, son of Omri, king of Judah.

Over the next year several consignments of large wooden cases laden with sculptured stone, ivory, colossal lions, and winged bulls floated on wooden rafts (borne on the same animal bladders depicted in the Assyrian sculptures in the cargo) down the Tigris to the marshlands of Basra, then aboard ship on the Persian Gulf, and eventually to England.

BEFORE the arrival of the Nimrud bounty at the British Museum, Rawlinson was already back at the Khurasan pass to copy the Babylonian glyphs. This script was remarkably similar to that engraved on the Assyrian monuments now reaching London and Paris; however, its five hundred different signs created a formidable barrier to decipherment. After three full years of unrelenting effort and baffling dead ends, a despondent Rawlinson conceded that "he had more than once been on the point of giving up because he had lost all hope of ever obtaining a satisfactory result."

In those phrases of the Old Babylonian text that contained the royal names and titles, Rawlinson noted that some words, such as those for *king* and *son,* were evidently expressed by single signs and not by a set of letters as was the case for Old Persian. This new insight suggested that ancient Babylonian was, in part, primitive and richly ideographic. Known names such as Xerxes had substantially fewer signs in the Babylonian version than in the Old Persian. If the language was spoken by a Semitic people, Rawlinson reasoned, the ancient scribes may have written only their consonants as was done with Hebrew. But when he took the consonants derived for one key word and applied them to others, he found that they allowed a large number of phonetic possibilities.

Just when it all seemed impenetrable, a new insight in 1851 pried apart the riddle. Rawlinson realized that Babylonian was written without simple consonant glyphs. Apparently the script had no individual letters at all—at least as we use them today in the Indo-European languages. Instead, the tongue was an "agglutinated" ensemble of syllabic sounds, each represented by a distinct sign. An individual glyph represented a combination of a vowel and a consonant. Finally, the Assyriologists were hearing phantom voices emanating from the stones of Mesopotamian monuments. The tongue they spoke became known formally as Akkadian. For a brief period it was thought to be the first written language in the world. For more than fifteen hundred years it had been the common dialect of "the land between the two rivers."

WHEN the final shipment departed Nimrud in 1847 with one of the best-preserved colossal bulls, Layard filled his trenches with earth for the last time and returned to Mosul. Earlier that year he had begun to probe the mound of Kuyunjik, the city of Mespila in Xenophon's time, circa 380 B.C. The ruins could be visited from Mosul by walking across a bridge over the Tigris. The most delicate issue was its active use as an Islamic shrine and cemetery.

Working inconspicuously at first, Layard encountered at the northernmost entrance gate the largest pair of winged creatures yet unearthed. They dwarfed those from Nimrud. Then, digging more boldly, he came upon a series of limestone slabs of a great palace. With Rawlinson's aid in translating their inscriptions he learned that the metropolis had been erected by Sennacherib (704–681 B.C.) and that this king's residence was indeed biblical Nineveh. Its masterfully carved bas-reliefs celebrated the siege and the successful Assyrian assault on a fortified enemy hilltop town. The panorama, filling several large panels, represented the famous attack on biblical Lachish, that was described in 2 Kings. The magnificent carvings now on permanent display in the British Museum chronicled a stunning defeat of the southern kingdom of Judah.

During his formal explorations at Nineveh, Layard focused his effort on the interior of the central mound. According to his monograph:

> In this magnificent edifice I had opened no less than seventy-one halls, chambers, and passages, whose walls, almost without exception, had been paneled with slabs of sculptured alabaster recording the wars, the triumphs, and the great deeds of the Assyrian king. By a rough calculation, about 9,880 feet, or nearly two miles, of bas-reliefs, with twenty-seven portals, formed by colossal winged bulls and lion-sphinxes, were uncovered in that part alone of the building explored during my researches.

These murals, some depicting in chronological order the construction of the city, with chain gangs of imprisoned slaves pulling sledges on rollers laden with the colossal bull sculptures, were themselves only a prelude. Tunneling down into the rubble with the light of torches, the workers tumbled through the ceiling of King Assurbanipal's royal library (the monarch had ruled from 668 to about 627 B.C.). The floors of two vaulted rooms lay strewn with rectangular sunbaked clay tablets. Many were broken, but, remarkably, a number were still intact. Every tablet appeared divided into multiple columns, and within each column row upon row of the wedge-shaped cuneiform characters greeted the excavator's eye. Layard's workers eventually packed up more than twenty-four thousand fragments and shipped them off to London. Here was the real treasure that Rawlinson had urged Layard to search for—the accumulated written wisdom of a lost civilization: its law, its medicine, its commerce, its literature, and its mythology. The fragments from one of these tablets would find their way to George Smith twenty-five years later.

T H E first public viewing of Layard's findings coincided with the 1851 Grand Exposition in London, in celebration of Victorian progress. Assyrian revival became the fashion of the day. Bracelets, pins, and silver-gilt caskets were designed and marketed in motifs copied from the monumental inscriptions of the Nimrud and Nineveh palaces. The riches from the Assyrian kingdoms restored interest in the places, people, and events of the Old Testament. Despite the setback to the biblical flood as a real global event brought about by William Buckland's abandonment of diluvial theory, the Assyrian genealogies, and a host of other unique details from the distant past— affirmed by the Nineveh treasures—nourished in the Christian laity a steadfast conviction that Scripture was indeed a true chronicle of antiquity.

CHAPTER FOUR

The Face of the Deep

THE tablets from King Assurbanipal's royal library in Nineveh did not disappoint the biblical scholars at the British Museum. Their attention focused immediately on the inscriptions of the monarchs whose exploits and reputations had also been described in the Old Testament. The aggrandizing palace propaganda of such prominent sovereigns as Tiglath-Pileser III, Shalmaneser V, Sargon II, Sennacherib, and Nebuchadnezzar II imparted a new historical authenticity to Scripture.

The invincibility of King Nebuchadnezzar II was legendary. His notoriety as a warrior began when, as crown prince of Babylon, he turned back the mammoth army of Egypt in a famous battle at Carchemish, with heavy casualties on both sides.

> . . . for the mighty man has stumbled against the mighty, and they are fallen both of them together. (Jeremiah 46:12)

Of course Nebuchadnezzar was even better known to the Bible reader for his eighteen-month-long siege of Jerusalem, the eventual breaching of its walls, the looting of the holy city—its temple torched and razed to the ground, its rulers put to death, and the surviving citizens of Judah deported into generations of bondage.

While the investigation of the Nineveh tablets continued at the British Museum, one of the most fervent believers in the accuracy of the Scripture, young George Smith, realized that the Chaldean scribes had copied their literature largely from much older sources. According to Smith, "It appeared likely that a search among the fragments of Assyrian inscriptions would yield traces at least of some of the ancient Babylonian legends." Smith's supervisor

George Smith pieces together fragments of a tablet containing
the story of the great flood

in the Department of Oriental Antiquities, Sir Henry Creswicke Rawlinson, pointed out "several coincidences between the geography of Babylonia and the account of Eden in Genesis, and . . . the great probability that the accounts in Genesis had a Babylonian origin."

Thus it was that Smith began cataloguing bits and pieces of Henry Layard's cuneiform tablets into a collection entitled the *mythological and mythical.* Prior to chancing upon and translating the curious partial tablet with the astonishing statement that "the ship rested on the mountains of Nizir, followed by the account of the sending forth of the dove, and its finding no resting place and returning," Smith had noticed a piece numbered K63 with a reference to a creation myth. However, in his excitement at discovering a predecessor to the biblical Noah, Smith temporarily set aside this fragment. Instead of following this new lead, he focused his attention on finding other missing elements to the flood story. He wrote in his diary at the time,

This search was a long and heavy work, for there were thousands of fragments to go over.

In a frenetic few weeks three separate and slightly different versions of the Deluge story coalesced from the *mythological and mythical* collection. None was complete, and each was assembled from a joining of just a few small fragments. Of the six columns of text in the original tablets, the compositions of the third and fourth were now nearly restored, but the others were either missing entirely or only partially readable due to gaping voids. Nevertheless, with resolve and integrity Smith was able to decipher enough of the essence of the story of the deluge to prepare his eagerly awaited lecture to the Society of Biblical Archaeology on December 3, 1872.

One morning in his windowless alcove he noticed that the glyphs signifying the hero in the deluge tablet reappeared on another remarkably intact tablet. The text there contained no mention of a flood. Instead, Smith read that his familiar hero dismissed a seductive marriage proposal from the goddess of love, and slew the "Bull of Heaven" sent to earth to avenge her humiliation.

At that moment Smith understood that the deluge tablet was one chapter in a much larger epic tale, the adventures of a king from the city of Erech who, when confronted with his own mortality after the death of his dearest friend, embarked on a long journey in search of eternal life. Although the story was "poetical and exaggerated," in Smith's words, he believed it contained "a basis of truth, and that this monarch really reigned and founded the Babylonian kingdom."

Smith's lecture to the Society of Biblical Archaeology stirred such emotion that the proprietors of the London *Daily Telegraph* offered a reward of one thousand guineas to reopen the excavations at Nineveh and "bring back the lost tablets." The trustees of the Museum wasted no time accepting the donation. They instructed Smith to make a short excavation and granted him a six-month leave of absence for the endeavor.

Although delayed for two months in Mosul before receiving permission to enter the tell of Kuyunjik and climb down the staircase into the royal library of King Assurbanipal, once inside Smith was almost instantly rewarded. He quickly sent a dispatch to the *Daily Telegraph* reporting that on the evening of his fifth day on site,

> I found a new fragment of the Chaldean account of the Deluge belonging to the first column of the tablet, relating the command to build and fill the ark.

Smith cabled the newspaper proprietors that the lost piece he was sent to find

. . . contained the greater portion of seventeen lines of inscription . . .
fitting into the only place were there was a serious blank in the story.

This verse described supplying the ship with food and placing in it treasures, seeds, animals, and the children of all the craftsmen. In the text sent
to London, Smith saw a striking difference between the inscription and the
Bible with regard to the nature of the ark. In the tablet from Nineveh the
vessel was a regular ship, navigated by mariners and launched into the sea.
To Smith, experienced in the Akkadian language, the vocabulary implied
that the deluge story was "the tradition of a seafaring people, or at least of a
people lying in the lowlands near the mouth of a great river."

In the weeks ahead Smith recovered additional chapters to the story,
eventually including portions from all twelve tablets of what we now know
as the Gilgamesh legend, of which the deluge was the eleventh tablet. Today
Gilgamesh is considered the most significant literary work to come out of
ancient Mesopotamia, composed and recomposed over a span of two thousand years, with the words kept alive in the traditions of the singing bard
long before they were committed to writing in the cuneiform glyphs. The
many different renditions and the multiplicity of language families, Sumerian, Semitic, and Indo-European, in which the story was passed down
through the generations attested that *Gilgamesh* attained the highest popu-
larity of all the literary masterpieces originating from Mesopotamia.

Smith's goal in traveling to Nineveh was to fill in the blanks of the "late"
or "standard" version. He believed that *Gilgamesh* had achieved the status
of "a national poem to the Babylonians, similar in some respects to those of
Homer among the Greeks." Like Homer's Achilles, whose warrior's pride
unleashed the tragedy of the *Iliad,* the arrogance of *Gilgamesh*'s champion
ultimately turned to grief and despair. Through the instrument of tragedy
the poet seeks to deliver a universal truth. The opening line of *Gilgamesh*
commands authority—extolling the great wisdom the hero gained from the
adventure about to unfold—for in that line "the waters of the fountain he
had seen."

Smith wrote back to London that his next good fortune

was a fragment evidently belonging to the creation of the world; this was
the upper corner of a tablet, and gave a fragmentary account of the creation of animals.

A few feet away in the same trench two more portions were dusted off,
"one giving the Creation and fall of man; the other having part of the

war between the gods and evil spirits." Though greatly impressed by the accuracy of Rawlinson's prediction of a Babylonian genesis myth far older than the one in the Hebrew Bible, Smith was distracted by the pressure of concluding his excavations. Only a few hours remained before he had to set out on his return to England.

Smith brought home hundreds of fragmentary clay tablets, each labeled with a unique number—in every instance prefixed with the letters DT in acknowledgment of the *Daily Telegraph* sponsorship. Perhaps due to its imperfect state of preservation because of the conflagration that destroyed Assurbanipal's palace, Smith did not immediately recognize the importance of the last fragment he gathered at Nineveh in May 1873.

THE opportunity to return to the Kuyunjik mound arose a year and a half later. This time the British Museum put up its own funds. Smith exploited the second occasion at Nineveh to advance his search for literary tablets not specifically related to the deluge. Since they had been composed of fine clay and inscribed in their wet state prior to baking in a furnace, most were extensively splintered. In some cases an individual tablet required painstaking restoration from more than one hundred separate slivers. However, the new effort brought to the forefront the connection between fragment K63 from Layard's excavations and the one Smith collected in haste at the end of his first expedition. In his field report, reprinted in the March 4, 1875, edition of the *Daily Telegraph*, Smith could hardly conceal his exhilaration when he announced:

> I [have] excavated . . . another portion belonging to this [Creation] story, far more precious—in fact, I think to the general public, the most interesting and remarkable cuneiform tablet yet discovered. This turns out to contain the story of man's original innocence, of the temptation, and of the fall.
>
> . . . The Story of the Creation and Fall, when complete . . . is much longer and fuller than the corresponding account in the *Book of Genesis*.
>
> . . . The narrative on the Assyrian tablets commences with a description of the period before the world was created, when there existed a chaos or confusion. The desolate and empty state of the universe and the generation by chaos of monsters are vividly given. The chaos is presided over by a female power named Tisalat and Tiamat.
>
> . . . [T]he discovery of this single relic in my opinion increases many times over the value of the *Daily Telegraph* collection.

Smith was struck by the uncanny similarity between the opening lines of the Babylonian creation myth and the initial two verses of the first chapter of Genesis, which he had memorized in boyhood:

1. In the beginning God created the heaven and the earth.
2. And the earth was without form and void; and the darkness was upon the face of the deep. And the spirit of God moved upon the face of the waters.

The cuneiform glyphs on the tablet he had just dusted off and was now reading spoke with an identical idiom:

1. When above, were not raised the heavens:
2. and below on earth a plant had not grown up;
3. the abyss also had not broken open their boundaries:
4. The chaos (or water) Tiamat (the sea) was the producing-mother of the whole of them.

Smith could barely trust his own decipherment. Both narratives not only shared a story line, but also incorporated the identical word, *chaos* (a primitive sea monster), to connote the watery deep—*Tehom* or *Tehomot* in Hebrew and *Tiamat* in Akkadian. In Smith's translation of the Babylonian creation account, the sea monster Tiamat is slain and her body is cut in two, with one half placed in the firmament to prevent her waters from ever flooding the earth. The saltwater terror is kept in abeyance for perpetuity. The exact same image, a bolt shot across a gate, is found in the Book of Job.

Other intriguing literary comparisons also became evident. The "Tree of Life" in the Garden of Eden was almost certainly the sacred grove of Anu (the ancient Mesopotamian sky god who was the prime mover in creation, and the distant, supreme leader of the gods), which the Nineveh tablet described as "guarded by a sword turning to all the four points of the compass." The divine tree was often embossed on the sculptures of the palaces at Nimrud and Nineveh as well as on cylinders used to make royal seals for temple documents. In pictorial representation the sacred tree was invariably accompanied by a serpent.

The Babylonian tablets thus raised for Smith and his peers, schooled in the Scripture, a ream of challenging questions: How could the texts from Mesopotamia have such striking parallels with Scripture, including the essence of their shared myth, the use of vocabulary with nearly the same pronunciation, and phrases worded almost identically? Was the Bible no longer the original historical source of human creation?

There was no doubt that the deluge described so vividly in the Gilgamesh legend had been inscribed on stone tablets long before the writing of the first books of the Old Testament. Might *Gilgamesh* be the original account of Noah's flood? If that was the case, then it had been carried by the Israelites as an oral tradition before its eventual redaction in Hebrew script.

In spite of the alternative heretical possibility of pure invention in both versions, Smith held fast to his personal belief in a factual basis for the ancient texts. He promised the inquiring readers of the *Daily Telegraph* that

> when my investigations are completed I will publish a full account and translation of these Genesis legends, all of which I have now been fortunate to find.

Tragically, Smith was unable to fulfill his pledge. On his third trip to Nineveh in 1876, he was detained in Istanbul for six months while trying to obtain his permit to excavate. Not heeding a warning about the danger of a summertime journey to Aleppo, he succumbed to a lethal virus contracted from a borrowed canteen when sweltering under the blazing desert sun. Born the very month that scientific belief in the universal deluge expired upon Buckland's conversion to the theory of the Ice Age, George Smith perished at age thirty-six.

CHAPTER FIVE

Ur of the Chaldees

EORGE Smith's legacy to future generations was the certainty that a flood, so overwhelming that it divided human history into a before and after, had indeed taken place. The flood story originated in Mesopotamia, so he reasoned that one should go there to look for its physical evidence. Indeed, the "Kings-Lists" tabulating the years of reign of royalty had separated those ruling in the antediluvian from those in the postdiluvian. Smith had repaired a tablet that mentioned one of the antediluvian kings by name. He was the "Man of Shuruppak," who was instructed to "tear down his house and build a ship." The inscription placed his city on "Euphrates' banks" at a site that archaeologists have since recognized as Tell Fara, about halfway between Babylon and the head of the Persian Gulf.

In his short life Smith defined our concept of the flood. Picking up Smith's trail of inquiry was Charles Leonard Woolley, the son of a clergyman. Following his education at New College in Oxford under the illustrious archaeologist Sir Arthur Evans, he became assistant director of Oxford's Ashmolean Museum and led his first excavations in Syria between 1911 and 1914. There he developed a knack for working effectively within the Ottoman political sphere. This skill served Britain well during World War I until his role as an intelligence officer was exposed, and he was captured and imprisoned.

After his repatriation, Woolley's moment of opportunity came in 1922 when he was appointed expedition leader of a joint British Museum and University of Pennsylvania dig at the Tell al-Muqayyar, some fifty miles south of Tell Fara in what is now southern Iraq. The cuneiform script embossed on previously unearthed foundation stones of this ruin had been translated seventy years earlier by none other than Henry Creswicke Rawlinson, who had astonished his peers when identifying this lost city as the Ur of the

*Charles Leonard Woolley and his wife Katherine examine the prehistory of
Mesopotamia amid excavations in the royal cemetery at Ur*

Chaldees, the birthplace of Abraham according to Genesis 11:28, 31. Woolley
worked the excavation for a dozen years, exposing broad swaths of its
former metropolis and revealing a vivid picture of Old Babylonian life late
in the third millennium B.C., around the time that the flood legend was
preserved in script. Within its residential areas he sorted through the remains
of buildings and courtyards, revealing the street plan and assigning to the
byways familiar names from his school days in Oxford—Broad Street,
Church Lane, and Paternoster Row. He found himself intrigued by how
closely the two-story brick houses of the distant past resembled those still in
use in the old quarters of the nearby town of An Nasiriya; their doors and
windows were in the same style of reed woven into wooden frames.

Having a strong belief in the historical and religious truths of the Bible,
Woolley became fascinated by additional king-lists unearthed in the course
of his excavations. He used them as the primary source of Ur's history,
taking particular note of those rulers who reigned before the flood and the
numerous dynasties that bridged the long period from the flood until Ur
rose to the splendor of a major metropolis.

But Woolley felt that he could not trust the chronology of the lists. Although the dynasties were consecutive, an individual ruler often had a life
span of hundreds of years. However, at different levels in his dig he would
uncover an inscription on a monument containing the name of one of these
monarchs, which strongly suggested that the person had been a real sover-

eign. Woolley asked, "Are the king-lists to be dismissed as mere fables?" And then, not waiting for others to answer, he replied, "We need not try to make history out of the legends, but we ought to assume that beneath much that is artificial or incredible there lurks something of fact."

What impressed Woolley profoundly was the listing of cities that had existed before the flood, implying an earlier people in Mesopotamia before the deluge. He reasoned, "The total destruction of the human race is of course not involved, nor is even the total destruction of the inhabitants of the delta . . . but enough damage could be done to make a landmark in history and to define an epoch."

Woolley's conjecture of previous inhabitants suggested that if he dug deep enough beneath Ur, he might encounter the flood deposit, its silt strewn across the land when the deltas of the Tigris and Euphrates were drowned. He hypothesized that this horizon would separate those who came before the flood from those who came after. Five years into his project, and with his crew of local laborers by then well trained in sorting through layers of debris, he sank deep shafts that led him to the royal cemeteries. When opened, the tombs displayed an interment of king and queen accompanied by human sacrifice on an extravagant scale. The entire retinue of the ancient court—its servants, soldiers of the guard, musicians, the ox that pulled the funeral cart, the driver and grooms—had been laid out as if they had fallen asleep under a spell at the foot of the wooden bier upon which the monarch lay, with the queen on her own platform wearing a floral headdress made of paper-thin leaves and flowers of gold, silver, and electrum. Every individual appeared poised to accompany the regent into the afterlife. To Woolley's experienced eye the grave objects expressed an art form and metallurgy so advanced that they could not have been achieved without a long period of gestation. Even the architecture was revolutionary in its use of the arch, vault, and dome—inventions that would not reappear outside of Mesopotamia until the time of the Romans almost forty centuries later.

Keeping to his plan Woolley dug deeper yet and eventually reached the very first settlers of southern Mesopotamia when Ur was little more than a marsh village. It was on his way through these strata that Woolley encountered the material evidence of a great inundation, represented in ten continuous feet of waterborne silt devoid of human artifacts. The silt blanketed houses and temples. Given the overall flatness of the surrounding countryside, Woolley saw the substantial thickness of this sterile deposit as evidence of a wide expanse of submergence. For him it was easy to believe that the bulk of the older population perished in a single event. He would write many years later about the region affected by the inundation: "With the land too valuable to be left unattended, a horde of immigrants flocked into it from the north, settling down with the few survivors and bringing with them new arts."

The discovery of what Woolley assumed to be an actual deposit from the flood in Genesis electrified the public—clergy and laity alike—as the news spread like wildfire around the globe in newspaper headlines, transoceanic radio broadcasts, and the newsreels delivered to movie houses. Not only had lost cities of the Bible risen from the desert sands, but the narratives in Scripture had also been confirmed. For several decades Woolley's interpretation of the silt layer further confirmed the Old Testament as a credible source of human history; the power of his persuasion was based more on enthusiasm and his gift for writing to the popular audience than on scholarly argument. Following its publication in 1929, his *Ur of the Chaldees* became the most widely read book on archaeology ever printed.

However, subsequent trenching at Ur, in the neighboring tells that surround Ur, such as Abu Shahrain (biblical Eridu), and in those extending north to other equally ancient settlements, such as Tell el Oueili and Choga Mami, have invariably failed to encounter this same silt layer. After much probing by trench and drill to trace its extent, investigators have determined that the surface area of the deposit was localized and perhaps only a single breach in a levee of the Euphrates River, forming what modern hydrologists call a "splay deposit," covering at most a few square miles of the lateral floodplain. No archaeologist today considers Woolley's silt layer at Ur to be any more significant than a thousand other silt layers spewed from the two great rivers during and since the last ice age. None of these local floods apparently had more importance than any other in serving as a major divide in human settlement in Mesopotamia.

As recently as the eighteenth century practically every scientist and intellectual curious about the natural world and its workings was driven to harmonize science with religion. In Britain's centers of learning, including Cambridge and Oxford, this blending was advanced to the extreme exemplified by the Reverend William Buckland. In studying nature, he said, he would be studying the unfolding of God's wise and benevolent plan. The great quantities of water needed for a universal deluge presented no problem. The flood happened in a world different from ours today; for example, when "the Face of the Earth before the deluge was smooth, regular, and uniform; without Mountains and without a Sea." To submerge all the land one would need no more water than today. The supply came from "the great Abysse, the Sea, or Subterraneous water hid in the bowels of the earth," or it rained down out of the sky during a close encounter with a comet.

Wild theories abounded, unconstrained by fact and fundamental scientific principles, until it began to be appreciated that the earth was indeed ancient and far older than the popular chronology of James Ussher, archbishop of Ireland, who in the reign of Charles I during the mid seventeenth century had calculated from the genealogies in Scripture that the heavens and the earth had been created on the evening before Sunday, October 23, 4004 B.C. The first geologist to provide a strong and convincing argument for the immense antiquity of an earth with "no vestige of a beginning—no prospect for an end" was James Hutton (1726–1797). His monumental "Theory of the Earth," published in the *Proceedings of the Royal Society,* left no room for the flood. Human presence on the planet occurred in the most recent and tiny fraction of its multibillion-year history, long after the appearance of fish, amphibians, reptiles, plants, birds, and mammals.

Hutton had in his day but a momentary impact on Western thought, gaining adherents within pockets of the Enlightenment on the continent of Europe. But Buckland was one among many historical geologists, nicknamed diluvialists, who viewed the earth as having experienced a series of "revolutions," each reshaping the continents and seas, extinguishing many of their former inhabitants, and making room for new ones. Noah's flood was the most recent of these annihilations and repopulations. The "drift deposits" of boulder-strewn clay scattered widely across Scotland and northern Europe served as the perfect manifestation of a great torrent of destruction. Even Charles Lyell (in his first edition of *Principles of Geology*) was initially unchecked in his enthusiasm for a flood origin of this residue, although he preferred to attribute the erratic boulders to debris dropped from icebergs set afloat by the rapidly rising ocean.

Belief in the flood as an agent of global change has subsided in the century since Louis Agassiz's momentous discovery of the Ice Age and the conversions of Buckland and Lyell previously described. A modern essay on the Genesis story in Western thought says the "diluvial theory came to be seen as what it was: one of the many imaginative but mistaken ventures that have accompanied the development of earth-science." Its remaining proponents are the Christian fundamentalists and the movement in the United States called Creationism.

ALTHOUGH the mythical flood has not been substantiated in the record of earth history, George Smith, Henry Creswicke Rawlinson, and Charles Leonard Woolley may not have been far from the mark. None of these gifted scholars thought the Mesopotamian legends implied a global submersion. Each saw the flood of legend as uniquely overpowering and affecting a people who were the ancestors to those who founded the great city-states of southern Mesopotamia. In searching for its physical characteristics, Rawlinson, Smith, and Woolley viewed the legends in the context of their own personal experiences: The flood derived from a storm bringing rain and perhaps a high tide on the landscape, especially in coastal regions. Though the downpour in myth had been exceptional in its destruction, when it was over, the waters receded, the land dried, and the cities sprang back to life.

But what if the myth story is examined in a different light, one that asks if human history was cleaved into a before and after primarily because the inundation was permanent and not temporary? Might a flood that never subsided have expelled a people from their former homeland and forced them to find a new place to live? Might the boat that the survivors built and filled with seed and animals have been not for the purpose of repopulating the earth but for transporting to safety the items vital to a continued subsistence (domesticated plants, animals, and knowledge of the arts)? And what if in addressing such a hypothesis one does not limit the investigation to Mesopotamia but looks beyond its borders?

The resources to look for a world hidden in the darkness of the abyss reach well beyond those available to the traditional archaeologist who opens tells and resurrects lost cities. Using the tools of the ocean sciences, climatology, radiocarbon dating, anthropology, genetics, and linguistics, a new group of explorers will again probe the prehistory unveiled by Rawlinson, Smith, Agassiz, Layard, and Woolley. The new journey will redirect the public once more to 1872 and, with an imagination stimulated during the relating of the adventures on the following pages, to a new meaning in the message delivered from clay fragments on that foggy morning in the dimly lit alcove of the British Museum.

The Discovery
of a Real Flood

CHAPTER SIX

Hidden River

THE equipment-cluttered research vessel *Chain* shuddered as its powerful twin screws began to churn the olive waters of the Golden Horn into swirling eddies of froth. To the scientists already at work inside the main laboratory of this floating observatory the rumbling meant that they were now under way. Looking up from a workbench along the port bulkhead, they could see the pilings of the wharf transferred from one porthole to the next as the eighteen-hundred-ton ship slowly backed away from its slip in a harbor choked with coal smoke. This day, October 18, 1961, would be historic for the Woods Hole Oceanographic Institution of Cape Cod, Massachusetts. Its flagship was about to undertake the first United States oceanographic venture into the Black Sea. Aboard were distinguished guests from the Turkish navy.

Out on the afterdeck in the dawn light the boatswain snaked the castoff hawsers into neat parallel rows where they would remain for a brief sixteen-hour adventure through the Bosporus Strait and into a sea world as practically unknown to the Americans as it had been to Jason and his Argonauts three thousand years earlier.

On the blue gantry that filled the aft deck, a technician had just completed his inspection of more than a hundred electronic thermometers attached along a cable of chain links wound onto a massive reel. This instrument string would soon be lowered into the water to measure the temperature variations of a feature called the thermocline, which bends sound as it travels through the oceans just as rays of light are spread apart when passing through a glass prism. For those who needed to find and track enemy submarines by listening to the churning of their propeller blades, the thermocline was of vital interest.

In the shrouded darkness of the previous night, a column of Soviet warships, betrayed by the sky-scraping silhouettes of their missile-guiding radar towers and their shrieking turbines, had steamed in the opposite direction of today's mission. The battleships and frigates had departed from Sevastopol and Novorossisk en route to the Atlantic and Indian oceans. The cold war was escalating and drawing the superpowers into a mutual show of force. A short while back NATO had supplied Turkey with nuclear-armed bombers. The Soviets were responding by projecting their own offensive might.

The *Chain*, commissioned in 1943 as an open-ocean salvage tug for the rescue of Allied shipping devastated by the German U-boat domination of the North Atlantic, was under contract to civilian scientists conducting U.S. military-sponsored oceanographic research. Its twenty-first expedition from home port in Woods Hole, across the Atlantic, and into the Mediterranean Sea had begun two months ago. Outfitted with the latest acoustic technology essential to ocean-floor mapping and antisubmarine warfare, the *Chain's* incursion, even with a Turkish presence, into the Black Sea and the territory of important Soviet naval bases might be viewed as provocative.

On the bridge, Captain Edmund Hiller relayed to his helmsman each order of the Turkish harbor pilot, to whom, by maritime convention, he had relinquished control of his ship for the passage through the congested and winding Bosporus Strait. Only upon reaching the open Black Sea would the skipper regain authority for directing its maneuvers. Hiller, standing erect in his dress uniform, was courteous to the foreign guests who climbed the stairwell from the mess deck, passed through his cramped chartroom, and ascended the steep ladder to the laboratory directly above the pilothouse. Nevertheless, his first mate could detect that his boss was tense about this last-minute interruption of a normal three-day port call for rest and relaxation. The mate and Hiller shared the same apprehension that the scientists soon would announce another surprise or two.

The small compartment above the pilothouse, known as the "top lab," served as the command post where rotating watches of technicians tended to the affairs of scientific data collection. Practically all the necessary electronic gear, in racks stretching from the deck to the overhead, was homemade, designed and fabricated in a sprawling pier-side machine shop back at Woods Hole. Even before the hawsers had been thrown off the dockside chocks, one of the scientists had fired up the echo sounder in its gray console that was nearly as tall and wide as the researchers themselves. The enormous bulk of this precision depth recorder was an indication of its vacuum tube heritage. The recorder's drum, at waist level, rotated at precisely four revolutions per second. With each turn a spiraling wire helix wiped moist electrolyte-treated paper against a thin steel band. As the contact between the helix and the paper began its trajectory along the band from left to right, a burst of stored electrical energy surged into a megaphone-like projector mounted on the keel of the ship beneath its bow. The powerful current expanded a ceramic crystal in the projector to create a pulse of sound that radiated into the water below. Projecting at nearly a mile per second and focused by the cylindrical shape of the crystal, the sound beamed directly at the bottom of the harbor.

As the *Chain* maneuvered into the Golden Horn, the vaulted dome of the sixth-century Hagia Sophia glided into sight. The oceanographers were awed by the sweeping historical panorama and the sun's rays reflecting from minarets and gold-domed mosques dating back to the Middle Ages. Little did the engineers and scientists know that in the course of the next several hours they would help illuminate an even earlier human past—a distant history when seas turned into deserts, forests vanished into tundra, and land was buried beneath ice; when our ancestors ceased hunting and gathering and began to harvest their first crops in the mythical Garden of Eden, and floodwaters broke loose upon the earth.

No sooner had the *Chain* slid away from the quay than a thin brown marking appeared at the left edge of the recording paper. It caught the eye of mission chief scientist Brackett Hersey, who could immediately recognize that this was the imprint of the sound bouncing back from the floor of the Golden Horn, a long estuary that had provided strategic protection to Constantinople through the millennia.

Meanwhile, naval officers decorated with shoulder braid, ribbons, and medals congregated in the top lab. Among them was the chief of the Turkish navy's hydrographic office. All the visitors were keen to get firsthand experience with these new sonar systems, especially in their own backyard. The plan of the day was to conduct a reconnaissance echo-sounding survey along the full seventeen-nautical-mile length of the Bosporus. The *Chain* would proceed on zigzag tacks that would take the vessel from a few hundred yards off the Asian shore to within a few hundred yards of the European shore.

The routes of ships traveling in narrow straits like the Bosporus are tightly controlled; traffic flows in separate lanes. Since the *Chain* would be cutting back and forth across these lanes during the mapping exercise, visual signals had been hoisted in the overhead rigging, and the shortwave radio was on,

The voyage of the Chain *up the Bosporus in 1961*

ready to announce each new course change. The whereabouts of dozens of other ships, small and large, absorbed the full attention of Hiller and his pilot.

In the top lab Hersey asked Bill Ryan, one of his junior assistants, to write the continuously changing depth beneath the ship on a large blank chart tacked to a tabletop, using the latitudes and longitudes derived from the compass bearings to locate each new sounding. Ryan had recently graduated from college and was embarking on a scientific career in oceanography through an apprenticeship as an electronics technician and watch stander. As hundreds and hundreds of numbers accumulated in discrete patterns along the path the ship was taking, Ryan began to see that where the Bosporus was wide and relatively straight, it contained a central channel.

On the paper record spewing from the echo sounder, the visitors were dazzled that they could actually recognize the profiles of tiny sand ripples on the channel bed in the dark waters below them. The echo duration, amplitude, and spacing implied the presence of both sand and gravel. Ryan, just getting his feet wet in geology, failed to recognize that their asymmetry indicated a migration of the sediment carpet in a northward direction toward the Black Sea.

When the *Chain* entered the serpentine narrows between the steep bluffs of Kandilli and Kanlica, where King Darius had led his Persian army across a floating bridge to subjugate the Scythians in 534 B.C., the depth profile unveiled an underwater gorge more than 350 feet deep. Here the echoes stretched into a pronounced reverberation. Hersey told his attentive guests that the ringing of the echoes meant that the channel cut through bedrock, whereas before the channel floor was covered with loose sediment rippled by underwater currents.

The signs of currents were not surprising. The Black Sea is a huge basin more than six thousand feet deep and fed by numerous large rivers. It receives considerably more freshwater from rain and river discharge than it loses by surface evaporation. Consequently, its excess freshwater is expelled to the Mediterranean. The Bosporus Strait and its downstream counterpart, the Dardanelles, at the western end of the Sea of Marmara, provide the only outlet from the Black Sea. In some parts of these straits the velocity of the outflowing water exceeds five knots (six statute miles per hour).

In heading northward toward the Black Sea the *Chain* bucked this swift flow through the narrow gorge. The Turkish chief hydrographer began to tell the Americans that as far back as the Persian invasions of Byzantium in the early seventh century, there had been knowledge of another current flowing in the opposite direction to the surface current and lying below it. He then went on to recount that when the Byzantine emperor Heraclius crushed the invaders from Asia Minor, he flung the torso and head of his foe separately into the Bosporus. The bloated torso floated and drifted south in the surface current. The head sank directly to the bottom. Months later teeth washed ashore in gravel bars to the north.

The chief hydrographer continued with the tale of Jason and the Argonauts, who in quest of the Golden Fleece had been the first Greeks to enter and explore the Black Sea. From the era of Hellenic colonization of the Black Sea in the late eighth to the early seventh century B.C. until the time of steam power, mariners had exploited the bottom-hugging counter-current by lowering into it stone-laden baskets that would drag their boats northward against the swift surface outflow. The first true scientist to mea-sure the strength of the undercurrent had been a twenty-one-year-old Italian named Luigi Ferdinando Marsilli. In 1680 he tied white-painted corks to a sounding line and watched the corks stream north as the lead-weighted line approached the bottom.

The chief hydrographer believed that the hidden bottom current might have been the legendary River Achereon, described to Jason as gushing from Hades into the mysterious sea that he would be the first to dare cross by sail and oar. In just the last half hour the *Chain* itself had passed between the ominous "clashing rocks" of the Jason myth. Soon the vista ahead opened into a vast green seascape, distinct from the familiar azure blue of the Mediterranean.

The next shift of watch standers took their turn in plotting the depths and drawing a contour map of the sea bottom. Hersey examined the map and was startled to see that the Bosporus channel continued onto the continental shelf of the Black Sea. He thought to himself that this countercurrent must be a veritable underwater river. Using the intercom, he requested that some-one track down Bud Knott and send him up.

Knott was caught off guard to see Hersey standing in front of the echo sounder with an impish grin on his face. Knott had expected the worst, that one of the instruments had malfunctioned and that he would spend the next few hours doing repair work. Relieved, he listened attentively as Hersey asked him, out of range of the others, if the "sparker" would work in the fresher water of the Black Sea. Hersey was referring to a new mapping tool that released the energy of nearly a quarter pound of exploding dynamite and could be used to probe the thickness of the seabed sediments. Knott smiled as he nodded. Because the salt content of Black Sea water is about half that of normal seawater, Knott had reduced the electrode spacing in the device to accommodate the lower electrical conductivity of the Black Sea's water.

Knott admired Hersey, who had raised precious funds from the Office of Naval Research to build this sparker, the most powerful echo sounder of its kind. Knott informed Hersey that he could have the instrument up and running in a half hour. Knott then heard his boss ask whether the sparker could be used simultaneously with the string of electronic thermometers, something not yet tried. It was an important question.

Down in the hold of the ship, Knott had built a protective metal cage the size of a bank vault with a locked entrance to keep people out for their own safety. Inside were thirty-two jerry can–sized capacitors to store an enormous electrical charge delivered from a generator in the ship's engine room. The cage also held a gimbaled table supporting vacuum tubes as big as bowling pins. These tubes were switching devices whose filaments were so sensitive that they had to be uncoupled from the rolling and pitching motion of the ship. First the tubes acted to direct electrons into the capacitors, causing these storage devices to swell visibly as the voltage peaked. Given a trigger, the tubes also served to instantaneously dump the entire charge into a fist-diameter cable that ran from the cage, up through a stuffing tube, and outside to a steel sled presently tied to the afterdeck. Adjacent to the sled lay one hundred feet of this cable recently coiled in neat figure eights.

During deployment, the sled was towed about five feet beneath the surface. Whenever the charge was dumped, the electrical current used the conductivity of the seawater to generate a spark between the two electrodes. This miniature lightning bolt vaporized a cavity of seawater the size of a basketball. The explosion produced a shock so substantial that sound from it not only bounced off the bottom of the sea but entered into and traveled through the seabed. A few weeks ago in the Tyrrhenian Sea, east of Sardinia, at a depth of almost ten thousand feet under water, the sound from this sparker had penetrated through more than three thousand feet of sediment cover before echoing off the underlying volcanic bedrock. Unfortunately, after a few hours of operation the sled had blown itself apart.

Although the sled was now rebuilt and the cavitation process that had led to its failure was better understood, Knott was unsure about the wisdom of using the sled simultaneously with the electronic thermometers, which were also towed. Knott and Hersey were apprehensive that the electrical discharge released by the vacuum tubes, instead of being retained in the sled, might leak through the seawater to the chain links, an excellent conducting material to which the thermometers were attached. Such a diversion could happen if the sled were again to self-destruct. The surge might fry the thermal sensors and possibly electrocute anyone near the large blue gantry on deck.

But the string of thermometers was crucial to measure the Black Sea's thermocline, whose first detailed imaging would be a major accomplishment. Furthermore, Hersey was anxious to use the sparker to find out whether the newly discovered channel on the Black Sea continental shelf was of recent origin and therefore sediment free or was an old relic feature and partly sediment filled. He phoned the bridge to ask Captain Hiller to come up. The skipper was responsible for all decisions that affected the well-being of the ship and her complement of crew, technicians, and scientists.

Hiller was accustomed to being placed in the awkward role of deciding what was permissible and what was not. He knew his mission was to accomplish the goals of the scientists who had raised the funds for all the operations, and important discoveries were made by new instruments that were little tested. But Hiller would be held responsible if something went wrong. Of course, when things went without a hitch, the scientists got all the credit.

Hersey and Hiller had sailed together on many prior expeditions and had developed a good working relationship. Hiller respected Hersey's scientific aims and proposed, in consultation with Knott, that the sparker sled be streamed from a long boom off the starboard side of the *Chain* to keep it away from the string of electronic thermometers that could only be towed from the stern. As a further precaution, Hiller and Hersey directed that the deck be cleared of all personnel for the duration of the survey.

After Hiller returned to the bridge, technicians gathered at the blue gantry, and the heavy torpedo-shaped weight at the bottom end of the thermometer string was soon hoisted from its stand, swung over the aft rail, and lowered into the opaque wake. The chain links clanked as they and the attached electronic sensors disappeared one by one into the foam.

Back in the top lab, another recorder was now displaying a hundred or more lines representing depths of equal temperature measured by the electronic thermometers. The thick lines indicated full degrees centigrade, and the thin lines showed the tenths of degrees. The surface temperature was a balmy 22 degrees centigrade (71 degrees Fahrenheit). Initially the temperature decreased only gradually with depth because the upper waters had been well stirred in the late summer and autumn storms. However, at a water depth of two hundred feet the display showed the temperature abruptly dropping to 8 degrees centigrade (45 degrees Fahrenheit) and remaining more or less uniformly cold the rest of the way to the seabed.

The *Chain* was now far out on the continental shelf where the water depth approached 250 feet. In a short while another ship appeared on the horizon. It had been picked up first by radar half an hour earlier at a distance of forty nautical miles and was closing range at thirty knots, far faster than the speed of commercial shipping.

The afterdeck was cleared as Knott prepared to launch the sparker sled. Only after this was accomplished could Hiller turn his attention to the new arrival. He had tried earlier to radio the ship, but no one had responded. Whoever they were, he knew they were not likely to be friendly.

Everybody felt the first explosion, a concussion from the expanding steam

The research vessel Chain *under the close scrutiny of a Soviet destroyer*

bubble of the sparker. The blast seemed especially powerful with the sled now towed abeam rather than behind the ship. The water surface above the sled lifted in a broad dome with each explosion, spray shooting skyward in jets and circular shock waves racing outward through the choppy sea. From the starboard wing of the bridge the sled appeared to be a darting porpoise, one moment gliding beneath the waves, the next moment vanishing under an expanding white cloud of steam.

Knott reported that sensors on the thermometer string were not picking up any stray electrical current, a good sign. The paper record of the echoes bouncing back from the explosions made by the sparker had begun to register as thin, wavy brown lines. These were reflections from layers upon layers of sediment long since buried by the steady rain of particles to the seabed. Hersey glanced at the scale settings that Knott had dialed in and gasped. He realized that through the medium of sound he was looking into the deep interior of the Black Sea's continental shelf, perhaps back to former seabeds millions of years old.

The *Chain,* shadowed and dwarfed by a tall Soviet destroyer, was now headed back toward the entry to Bosporus for the return trip to Istanbul. The escort had closed range to a ship length, and both vessels were side by side, separated only by the jets of water that were periodically rocketing skyward. There was still no contact by radio, no visual signal of any kind, not even curious Soviet sailors lining the rails. All that could be discerned on the destroyer were half a dozen officers on the port bridge extension, each with a pair of large binoculars.

Although clearly pursued, the *Chain* was in no apparent hurry. Above a speed of a few knots, water turbulence would drown out the echoes from the sparker. The research vessel steadied on a course to return to the channel that Hersey had realized earlier might extend from the Bosporus to the shelf edge. The ship's arrival at the channel was announced not only by the abrupt increase in water depth but by a complete change in the configuration of the bottom. Hersey and the Turkish chief hydrographer studied the profile and concluded that the channel was cut deep into the subsurface and only slightly filled with young sedimentary materials not found elsewhere. The two observers compared notes and measurements. It seemed that the channel, whatever its origin, was once a bone-clean cleft. Today it was being slowly buried by sediment carried in the current that was flowing through it.

In the excitement of this discovery, the technicians of the afternoon watch had completely ignored the piled-up paper recordings produced by the output of the thermometer string. The isothermal lines were now spread out across the whole width of the paper. Hersey looked confused as he discovered that the temperature just above the seabed was almost as warm as the sea surface. But that was impossible; warm water is lighter than cold water and always floats on the top. What could possibly keep it near the bottom? Was the thermister array malfunctioning? Could an electrical discharge from the sparker be interfering?

Hersey quickly became alarmed. Perhaps the sled was starting to tear apart, with welds cracking open from the repeated explosions. He turned urgently to Knott, who had already sensed that something was wrong. Before Hersey uttered a word, Knott threw open the circuit breaker to shut down the electric current. The reverberations of the last explosion died away.

Looking out the top lab window Hersey saw that the silent destroyer had moved directly astern of the *Chain*. His face became ashen. Had the Russians come too close, tangled in the steel links that carried the temperature sensors, and brought the bottom end of the string back to the surface? Or, worse, had they intentionally snared it? Hersey rushed to the intercom and called the pilothouse to inquire whether there had been any slowing to indicate that the destroyer was pulling on his gear. The helmsman reported no speed change. By this time several of the Turkish navy officers realized something was amiss.

At that moment the Turkish chief hydrographer peered up from the temperature recorder and announced that the sensor at the bottom of the string was measuring 15.5 degrees centigrade, a value he expected to find close to the floor of the channel. He then described how the countercurrent that they had first heard about back in the Bosporus consisted of warm, salty water from the Mediterranean. The salt, at a concentration twice that of Black Sea surface water, was heavy. That is why the warm water hugged the bottom in the channel. The heavier, saltier water from the Mediterranean was cascading into the Black Sea and displacing the lighter, fresher water that flowed in the opposite direction.

Hersey, relieved by this news, looked out the window. The Russian destroyer had slipped back a mile behind the *Chain*. It was bearing off to the east, having abandoned its inspection.

However, at the slow speed needed for the sparker operation, the *Chain* would never make it back to Istanbul before sunset. There was no point in turning the sparker back on, but the string of electronic thermometers could stay in the water and continue to operate at full speed. Thus, with the echo sounder pinging away and with diligent piloting to stay in the channel, the *Chain* followed the axis of the warm, salty current back through the strait and upstream into the gorge crossed by King Darius.

Hersey was unable to hide his amazement that directly beneath him a hidden river was literally pouring into the Black Sea. The chief hydrographer told an enraptured top lab audience how in September 1932 one of the small survey boats of the Turkish Hydrographic Service was towed through the Bosporus toward the Black Sea by a single bottle no larger than a vase, which had been designed to collect water samples and was lowered by wire into the core of this river.

THE brief outing for Americans to show off some new technology to NATO partners had evolved into a lesson in oceanography, geology, and history. Day after day through the ages an underwater river flowed north from the Mediterranean into the Black Sea. The echo soundings made by the sparker indicated that at some distant time in the past this river had been much more violent. It had scoured a sinuous cleft into solid bedrock and had sculpted cataracts and plunge pools that deepened toward the Black Sea. However, the strength of the current had waned through the millennia, allowing the channel to begin to fill with sediment. Borings into these infilling materials near Istanbul, carried out decades later in connection with a project to excavate a subway tunnel from Europe to Asia, would recover a body of debris in the very bottom of the chasm that contained a chaotic mixture of giant blocks of the bedrock, boulders, cobbles, sand, and seashells. The future researchers would find among these shells species of mollusks adapted to salty water that had come from the Mediterranean. At one time this stream had flowed north into the Black Sea in an extraordinary torrent, literally ripping apart the passage through which it coursed. What had unleashed such fury?

CHAPTER SEVEN

Gibraltar's Waterfall

THE needle on the large dial jumped and then settled back to a steady reading of 120,000 pounds. Then it jumped again. The tool pusher turned his head in the direction of the drilling superintendent who had also been watching the gauge. With a nod of his head the superintendent communicated his awareness of the sudden change in the formation. The tool pusher increased the pump pressure just a bit. The needle twitched and then began to bounce in steady pulses. He lowered the long arm of the lever used to regulate the rate of descent of the drill pipe. Screeches and hissing emanated from the huge winch behind the drill shack.

Bill Ryan, crouching behind and looking over the shoulder of the tool pusher, made an entry into his logbook under the heading of August 23, 1970:

> Core 13-122-4
> Started cutting at 19:45 hr
> 2328 meters below rig floor
> Hit something hard!

Ryan shifted his attention to the rotating pipe now slowed to a crawl in its descent through the center of the rig floor. Beyond he saw the orange orb of the setting sun intercept the flat cloudless horizon. Together they rose and fell through the steel-girded lattice of the derrick. In the gradually darkening sky overhead, a crescent moon swung back and forth, eclipsed at the zenith of its path by the hydraulic drill motor suspended in the two-hundred-foot-high tower.

A chilling breeze whipped whitecaps across the Balearic Sea. Sixty miles

The Glomar Challenger *drills into the bed of the Balearic Sea
to reveal secrets of the earth's distant past*

offshore of Barcelona, Spain, the drill ship *Glomar Challenger* rolled gently
on the growing swell. Every few seconds the ship's thrusters sprang to life
by command from a computer. In open tunnels that ran through the submerged hull at the bow and stern, an array of propellers churned the azure
water into lateral jets needed to fight the wind and current and keep the
floating scientific observatory at a single spot, 7,083 feet above the ocean
floor.

Ryan, almost nine years after his voyage through the Bosporus aboard
the *Chain,* was on another scientific expedition. This time he was co-chief
scientist on the thirteenth mission of the Deep Sea Drilling Project. Probing
into the sedimentary layer of the world's oceans had become the centerpiece

of an ambitious new program of the U.S. National Science Foundation. Its objective was to seek answers concerning the history and life of our planet. But instead of searching for this history in the strata of rocks on land, the contemporary adventure was being played out beneath the sea.

The sedimentary layer of the ocean floor is a virtual library of information. Its acquisitions are made of mud, ooze, sand, and rock, with a script choreographed by climate during the erosion of mountains by precipitation, the transport of dust from deserts by wind, the wandering of the sea in meanders and gyres, the freeze and thaw of ice caps, and the life spans of a multitude of creatures who once colonized the ground, both wet and dry, and left behind their skeletal remains in the form of fossils. The texts in this ocean floor library are deciphered by scientists exploiting the knowledge of physics, chemistry, and biology.

Earth's story is thus set in print as the acquisition of sedimentary particles by the seabed over the span of millions and millions of years, building layer upon layer of the seabed carpet, eventually reaching thousands of feet in thickness. The tale varies from place to place and from one era to the next. Some oceans have been bathed in warm tropical settings, while others have spent most of their existence at the frigid polar extreme. Occasionally a storm may rage through an ocean basin, stirring the accumulated sediments back into suspension. During such an event, a page or even an entire chapter of the ocean's history might be erased.

With enough drilling into the floor of the abyss, however, a remarkable account of the changing configuration of continents and oceans had begun to unravel site by site as the *Glomar Challenger* traversed first the Atlantic and then the Pacific Ocean, beginning its maiden voyage in 1968. Careful planning was crucial in order to place the drill holes where the sediment record would be best preserved and of the highest fidelity. Yet surprises happened. Nowhere did serendipity play a bigger role than in the first drilling of the Mediterranean Sea during the summer of 1970.

The rasping of the drum brake momentarily stopped. The tool pusher pointed out to the drilling supervisor an anomaly in the pump pressure. After a brief discussion the tool pusher engaged the massive winch. Its hydraulic motor screamed in protest as it hoisted the drill string to the top of the derrick. Shortly the drill pipe would be clamped in a circular vise called the Kelly table, and a joint would be unscrewed by the roughnecks with their giant air-powered wrenches called tongs. Next a cable would be inserted into the threaded throat of the open pipe and lowered to latch onto and retrieve the core barrel with its cargo of sediment just collected 535 feet below the seabed.

The Glomar Challenger *drilling the floor of the Mediterranean Sea in 1970*

Ryan left the driller's shack on the rig floor and walked down the narrow catwalk to the core lab. Two sedimentologists on the noon-to-midnight shift were leaning over a workbench in the center of the lab, absorbed in examining the core collected just before dinner. One was inserting a finger-sized plastic sampling tube into the moist and soft half cylinder of olive gray ooze resting in its storage tray. The other asked when they could expect the next core. Ryan replied, "In less than half an hour, and you can count on something different."

Ryan, thirty years old and the youngest member of the scientific team, washed the grime of the driller's shack from his hands and sat down at a binocular microscope to examine the ooze. As he peered at the chambered shells of tiny protozoa no larger than the head of a pin, he recognized creatures that had thrived in warm waters of the western Mediterranean more than four million years ago. The relative abundance of the different species, all belonging to the order Foraminifera, gave a measure of past climate going back millions of years.

Every now and then the cores contained a thin layer of sandy pumice, once blown high into the stratosphere by an exploding volcano of the type that had buried the cities of Pompeii and Herculaneum in A.D. 79. The pumice from Mount Vesuvius had been entombed forever in the deep sea. The first sediment core that Ryan ever examined when training to become an oceanographer contained pumice from Vesuvius that had rained out of the sky into the Bay of Naples, and when disturbed by earthquakes accompanying the eruption, had traveled in an underwater flow of sand and gravel for more than one hundred miles through a submarine canyon.

Ryan heard his name paged on the intercom. He was wanted right away on the rig floor. When he arrived, he discovered a serious problem. It appeared that the core barrel was stuck in the bottom of the drill string. The drilling supervisor suspected that the hole had caved in, that liquefied sand had flowed back into the drill pipe, and that this sand had wedged the core barrel against the inside wall of the drill pipe. That could be why the circulation pumps had overloaded and shut down. Whatever had flowed into the pipe had completely clogged its throat, making it impossible to continue drilling and coring. To Ryan the news was devastating.

Ryan headed for the bridge. Captain Joe Clarke had been apprised of the superintendent's decision to abandon the hole before Ryan arrived. It was the captain's unpleasant task to tell Ryan that his team needed to pick another site, one where unstable seabed conditions did not exist. The plugging of the drill pipe was a warning. In oil drilling it often forecast a zone where crude oil or natural gas resided under very high pressures. The *Glo-*

mar Challenger did not have a sufficient supply of heavy drill mud in its tanks to pump into the hole and control a sediment charged with hydrocarbons. A blowout at sea would be an environmental disaster of frightening proportions.

In the trials of the past two weeks Clarke and Ryan had built a relationship of mutual respect for each other's professional competence. Both were indefatigable scrappers of Irish heritage, and both took seriously their respective responsibilities: one for the safety and effectiveness of the mission, the other to maximize the scientific results. Their strategy was to make their separate charges coincide.

Three decks below the pilothouse, Ryan found Ken Hsü, co-leader of the expedition, in their stateroom completing his revisions of the hundred-page preliminary report of the previous drill site. With a compassionate embrace, Hsü shared the disappointment of his new friend. Born in China, Hsü had received his graduate-level education at the University of California at Riverside. Professionally trained at Shell Oil Company, he was now a world-renowned professor of sedimentology at the Swiss Technical University in Zurich. This expedition was his second voyage on the *Glomar Challenger*.

Hsü proceeded to pull tubes of charts from a bin beneath the double-decker bunk and unroll the papers onto the carpeted floor, the only available space in their cramped quarters. On hands and knees the two expedition leaders selected the next site, fourteen miles to the northeast and over the flank of a dormant volcano that protruded through a gap in the layer of sediment that had just clogged the drill string. During the few days that it would take to complete the sampling in that new hole, they could evaluate more calmly the circumstances that had just plugged the drill string. They hoped to be able to come up with a scheme to safely drill deep into the sediment layer that had thwarted their first attempt.

Ryan looked to Hsü for guidance in the analysis of the cored materials. Hsü depended on Ryan's knowledge of the Mediterranean seabed for interpretation of the tectonic history of this complex seaway that was being crushed during the past twenty-five million years between the converging African and European continents. Since Hsü would normally relieve Ryan in the drill shack at midnight, he offered to deliver the coordinates of the next drilling location to Clarke so that Ryan could get some sleep.

Wakened by Hsü in the early hours of the morning, Ryan found all the scientific laboratories abandoned. On the brightly lit rig floor, two roughnecks had just unscrewed the toothed drill bit. One of them put a bucket into Ryan's hand, remarking that it contained the culprit responsible for the hole collapse. The drill crew had used a fire hose to jet a forceful stream into the bottom strand of the drill pipe to remove the plug and free the core barrel.

With time to spare before dawn and a hot breakfast, Ryan returned to the core lab and emptied the contents of the bucket into the sink. Under a gently running faucet, he washed mud from pea-sized fragments of what appeared to be rock. Soon he noticed that there were only three kinds of rock. One was black with tiny cavities formed by bubbles once charged with steam in an erupting volcanic lava called basalt. Another was light brown and effervesced with drops of hydrochloric acid, indicating limestone. The third was transparent and contained elongated crystals with flat facets that resembled rock candy. The crystals were easily scratched and did not fizz with acid. As Ryan continued to wash the gravel, he encountered miniature seashells, no larger than the eraser of a pencil. Many were delicate, yet unbroken. More significantly, none were fragments of the larger shells one typically finds at the seashore.

Ryan selected about twenty specimens of each type of rock and several of each species of seashell and attached them with drops of instant glue to the inside pages of a manila folder. He then studied his collection under the binocular microscope. Bewildered by what he saw, he went back to the rig floor for more material. Every sample he washed consisted of the same three types of rock and the small seashells. He was unable to find anything else.

Trained at the Lamont-Doherty Geological Observatory of New York's Columbia University by two giants in the field of marine geology, Bruce Heezen and Maurice Ewing, Ryan was not surprised to have encountered gravel. It was commonly carried in mud slurries for hundreds of miles down the continental slope and across the continental rise to the abyssal plain by underwater landslides called "turbidity currents." After all, the *Glomar Challenger* had just been drilling into the floor of a canyon where gravel might be expected. Shells of shallow-water creatures a mile beneath the sea surface were not a puzzle in themselves, for they could also have been transported in the avalanches similar to those that had carried the Vesuvius pumice from the Bay of Naples deep into the sea.

What astonished Ryan was not so much the types of rock fragments present but the types that were absent. Gravel transported to the deep sea by "turbidity currents" should represent the erosional debris of the continents, carried by streams and rivers cascading out of mountains and across floodplains to the coast. The gravel should contain a large variety of continental rocks, with fragments of granite, gneiss, and graywacke (sedimentary rock containing the particles of even older mountain chains) that possessed lots of quartz, mica, and feldspar as mineral grains. Everything that would have been expected was missing. In its place were fragments of oceanic bedrock (black basalt), of hardened marine sediments (limestone), and the novel transparent crystals.

The sun was rising when Ryan, looking for company, located Maria Cita in the paleontology lab, deep in the interior of the drill ship. He showed her the collection he had prepared, and wanting to get her objective opinion, he said nothing except that it was the gravel formation that had clogged the drill pipe. Cita was upset that nobody had informed her the site had been abandoned. She told Ryan that the tiny seashells were dwarfed and fully mature specimens, about a tenth the size of normal adults. If he had brought them to her earlier, he would not have wasted his time looking for larger ones.

Cita, professor of stratigraphy at the Institute of Geology of the University of Milan in Italy, lectured to her attentive pupil about dwarfism, explaining that it was the consequence of natural selection in a highly stressed environment. As a micropaleontologist, her job was to identify the shells of plankton sieved from the sediment and use them to tell the age of the deposits and the nature of earlier environments. On the wall were diagrams that looked like the branches of a tree. The limbs traced the first evolutionary appearance and the eventual terminal extinction of these small animals through the past sixty-five million years.

Cita's desk was strewn with several monographs opened to pages filled with pen-and-ink sketches of the coiled shells. Some of the more recently published treatises contained exquisite portraits obtained by an electron microscope.

About this time Hsü appeared. His eye immediately caught the glitter of the shiny crystals resembling rock candy. "Selenite," he shouted. "From where?"

"From the gravel," Ryan replied. "My notes from the driller's shack indicate that we hit a few thin hard layers before the hole collapsed. Gravel could explain the bouncing of the drill bit. This collection represents its composition. The entire bottom of the pipe was full of it."

As Ryan confessed his bewilderment at finding only the dark volcanic pebbles, the brown limestone, the transparent crystals of selenite, and the dwarf seashells glued to the folder, Cita and Hsü became caught up in the puzzle. Hsü remarked that selenite was a crystalline form of gypsum, a mineral precipitated from water upon evaporation. Cita, familiar with gypsum-bearing strata widely distributed in the Apennine Mountains of Italy and Sicily, and known locally as the *Gessoso Solfifera,* commented that concentrated brine pools in which the gypsum formed would have been ideal settings for adaptation of seashore fauna to dwarf size. She pointed out that all the mollusks in the gravel indicated a coastal environment. Some species pointed to the beach zone, others to a lagoon behind a barrier spit. Cita even recognized some brackish water creatures that looked like specimens living today in the landlocked Caspian Sea.

By now Ryan was quite confused. He had expected Cita and Hsü to concur that the rock fragments and shells had been swept to the floor of the deep sea in watery avalanches. Instead it seemed that they thought these rocks had formed and the dwarfed creatures had lived right where they had been cored.

Shallow lagoons on the floor of the Mediterranean Sea? "No way!" Ryan protested. To him the idea seemed absurd even as a preliminary working hypothesis.

Hsü challenged his friend to deny the evidence assembled so neatly on the manila folder in the last few hours. If underwater landslides had been at work, where were all the other rock types that one would expect from the erosion of Spain and France by rivers? Still mystified and now hungry, the three chose to continue their discussion in the ship's mess. They were joined by Herb Stradner, an Austrian nannofossil expert. When he learned what they were so agitated about, he inquired about the age of the gravel. Offhand, Cita was unprepared to give a precise answer, because the sea shells were from species that lived unchanged over a long time, some still in existence. However, she had assigned to the previous core, cut above the gravel and thus younger, an age corresponding to the Lower Pliocene Epoch, an interval dated by a radiometric clock to between three and five million years ago. At the mention of the chips of brown limestone, Stradner asked if he might pulverize some of the fragments so he could view the powder through his more powerful microscope at a magnification of one thousand times.

The group returned to the paleontology lab, where, in a short time, Stradner prepared a dozen separate samples from different fragments of the limestone. The dust from each fragment was sprinkled on a glass slide and immersed in a drop of oil with special optical properties. He then brought the minuscule particles into various degrees of focus within the optics of his scope. They were the skeletal remains of tiny marine plants. As he rotated each slide in the stage of his microscope, Stradner noted a flickering in the illumination, which in turn gave him clues to the shape of the particles and hence the identification of the plants. He started to call out the names of extinct species he could identify with confidence. Referring to one of the wall charts, Cita located the particular branch with that species' name and translated its position on the evolutionary tree to an era in the past when this plant flourished in the sun-lit surface waters of the ancient Mediterranean. Stradner and Cita discovered that the limestone pieces belonged to the Miocene Epoch of the Cenozoic era and lived seven million years ago. This age indicated that the selenite and tiny seashells were sandwiched in time somewhere between seven and five million years ago. That was the same age as the Gessoso Solfifera. The team was dumbfounded by this coincidence. Hsü broke the silence. "Do you think the entire Mediterranean might have once dried up?"

Hsü volunteered to try out the idea on the rest of the shipboard scientific party. When the others greeted his suggestion with skepticism and even ridicule, Hsü had to admit that the evidence, a handful of gravel, was flimsy. Although Ryan was intrigued, privately he was uncomfortable with so preposterous an idea. Hsü and Cita were less inhibited. They wrote in their preliminary report, "We are stymied in offering an alternative, plausible explanation and urged Ryan to come up with a new site where the seabed could be safely penetrated." That required the *Glomar Challenger* to drill without any likelihood of intercepting more gravel and risking another collapsed hole. The scientists needed to make a strong case to Captain Clarke that they had an acceptable alternative plan.

Such an opportunity presented itself two days later when the *Glomar Challenger* finished drilling the dead volcano and set off for a location southeast of the island of Majorca. In the previous month the French oceanographic ship *Jean Charcot* had charted the Balearic Sea with equipment that looked into its sediment cover using sound waves. As the ship crisscrossed the margin of the central basin, it generated a network of sub-bottom reflection profiles from which the internal strata could be viewed and mapped. The survey exposed a region completely devoid of submarine canyons and the gravel they might harbor. The sediment layering appeared very uniform and without the fold structures that tend to trap oil and gas. Strata five million to seven million years old were in easy reach of the drill bit. The risk of encountering an overpressurized formation was minimal. The operations manager endorsed the drilling plan and forwarded it to Captain Clarke for approval.

In the middle of the night on August 27 the entire drill ship began to shake. Already the new hole was 1,270 feet into the seabed. Earlier drilling had proceeded smoothly and without event. Seven cores had been recovered and examined. The pounding increased in intensity and continued without abatement until sunrise. That could only mean that the drill bit was grinding on solid rock. From the French reflection profiles Ryan and Hsü knew that it was much too soon to be an encounter with the bedrock. However, in six hours the drill gained only nine feet of additional penetration. Hsü, impatient and fearful of a mishap, decided to bring up whatever had survived the thrashing.

A boisterous crowd gathered on the catwalk to watch the sand line hoist the steel core barrel out of the drill pipe. As the roughnecks extruded the transparent plastic liner and carried the tube with its cargo of rock to the awaiting scientists, someone shouted, "By God, a pillar from Atlantis!" Through the grease-smeared wall of the jacket, one could see that the white rock column was marbled in texture.

Out in the open on the worktable in the core lab, the five-foot cylinder of rock, only three inches in diameter, broke apart into a number of individual pieces. The scientists examined each fragment with their hand lenses. The thin laminations led someone to utter the word *Balatino,* a local Sicilian name for paper-thin layers of alabaster, a relatively soft material used by the Assyrians to carve the bas-reliefs for the palaces at Nimrud and Nineveh. One of the scientists held out a specimen for all to see. The light gray rock with dark streaks had the texture of hardened granular sugar, peppered through and through with pure white nodules. It was a typical component of the Gessoso Solfifera in Sicily.

The pieces in the top and bottom of the core tray resembled stacked candy wafers. In the middle·the layering was crinkled. One of the scientists volunteered to cut open several of the separate pieces with a diamond saw blade to reveal their internal structure. Hsü put a slice to his tongue and winced. He announced, "You'd better believe us now when we tell you that the Mediterranean dried up!"

What his tongue had identified was anhydrite, an especially dry variety of selenite that mineralizes under temperatures exceeding 110 degrees Fahrenheit. Anhydrite forms now only in the tidal flats of the extremely hot and arid Persian Gulf. These coastal lowlands, called *sabkhas* by the local Arabs, flood every decade or so in major storms. The rest of the time the sun beats down on them and evaporates any moisture it can draw up from the underground water table. Anhydrite is the precipitated residue left behind.

Hsü motioned at the wavy structures. "These," he pointed out, "are made by algae that secrete enormous organic mats over the moist sabkhas after each storm. Look, you can even see the outlines of their original cells. They require sunlight for photosynthesis!"

For two more days the *Glomar Challenger* pounded relentlessly into another hundred feet of the anhydrite rock until the drill bit wore out and the downward progress halted. Seven additional cores were received on deck with excitement. One by one they added further evidence, fortifying the emerging picture of a five- to seven-million-year-old Mediterranean desert landscape with drying lakes and their coastal mudflats, evaporating under a scorching sun.

As the *Glomar Challenger* sailed east into the Ionian Sea, between Sicily and Greece, Hsü, Cita, and Ryan speculated about the configuration of the desiccating Mediterranean. Had it been a montage of separate shallow lagoons interconnected to a marine waterway fed by Atlantic water entering through the Pillars of Hercules at Gibraltar, or had the entire region been cut off from the Atlantic by a natural dam with only trickles of water intermittently cascading down cataracts into an inferno of heat twenty to thirty times farther below sea level than Death Valley in California?

At the next site, southwest of the Peloponnesus, the arriving cores soon offered the first substantial clue about the depth of the seabed at the end of the desert phase. The scientists on the *Glomar Challenger* took the extra time and effort to drill two holes side by side so as to be confident that the cores had pushed into the sediment layer directly above the anhydrite deposits of the dried salt lake bed. These cores would contain a fairly complete record of what happened once the sea returned to flood the desert landscape. After the orange-brown mottled ooze was processed for its composition, faunal and floral content, and age, the team concluded in the expedition's initial reports that the ooze deposited right after "the growth of anhydrite in tidal flats is no different from that accumulating today." According to Cita, who used the bottom-dwelling fauna as a measure of water depth, the site of the dried-out salt lake had suddenly been transformed into a new marine sea thousands of feet deep and far removed from land. No transition was observed. The seabed creatures living directly on the anhydrite once the seawater returned were indicative of a "bathyal realm"—a term the paleoecologists generally reserved for the cold dark internal ocean three thousand or more feet below the warm sunlit sea surface. This observation supported the idea that the deserts had formed in a depression that had dried out and that the deserts had drowned suddenly by flooding of the depression under thousands of feet of newly supplied seawater.

Although Hsü, Cita, and Ryan viewed the new evidence as favoring depressions already in existence before the flooding, the other scientists were not entirely convinced. They suggested that instead of a rapid deluge to fill a deep hole with seawater, a shallow seabed may have subsided to make room for the water. In the absence of a consensus, the scientists agreed to engage in more drilling to evaluate the alternatives. However, first the advocates of each hypothesis had to formulate a set of predictions keyed to the essence of their preferred scenario. The hypothesis that failed to forecast adequately what they would find in the final weeks of the expedition would be discarded in lieu of the one that succeeded, a classic scientific procedure.

If a formerly shallow Mediterranean had sunk to become a deep basin, then the sediments accumulating after the anhydrite should preserve evidence of a progressive deepening of the sea bottom. One should see a transformation of formerly knee-deep lagoons to a multithousand-foot-deep sea taking place over a significant span of time. Since the fauna and flora in the sediment are relatively sensitive recorders of the decreasing temperature, increasing pressure, reduced food and oxygen and the rapidly diminishing light that all occur in a deepening ocean, such gradual sinking should have left behind a record for decipherment.

On the other hand, if the Mediterranean had always been deep except when it had dried out, then an abrupt drowning of its desert floor would have occurred once the floodgates opened. With the large hydraulic head of the Atlantic Ocean, there would have been no way to stop a deluge of salt water pouring through the proposed sluice gate at Gibraltar or wherever it might be.

The showdown between the "gradual sinking" proponents and the defenders of "instantaneous flooding" took place in the final weeks of the expedition as the *Glomar Challenger* entered the Tyrrhenian Sea to drill a site about seventy miles east of Sardinia. This is one of the youngest regions of the Mediterranean. Thus for proponents of a gradual sinking scenario, this next drill hole was ideally situated to reveal a progressive drowning of a former shallow platform.

In a water depth of ten thousand feet, the entire thickness of ooze above the anhydrite and alabaster was recovered in the cores. The drillers succeeded in recovering intact the actual contact between the evaporite rocks and the overlying marine ooze. Consequently, the cores brought on deck and then into the lab left no room for ambiguity. At that contact the passage from dry desert sand to the marine ooze was razor thin. The very first ooze on top of the windswept sand had been deposited fully in the "bathyal realm." Since such ooze typically accumulates at about an inch every thousand years, the abrupt change was too sudden to have lasted more than a century or as the environment changed from dry salt lake bed to a mile-deep abyss.

In the few days remaining before it was time to return to port, the drill ship returned to the Balearic Sea for a series of holes down the western continental slope of Sardinia and onto its abyssal plain. There the core samples produced rounded cobblestones and pebbles from an ancient braided streambed that had once coursed down the slope during rare rainstorms. These fluvial deposits were intermingled with a dry rusty-red soil still containing the seeds and fossil roots of desert sagebrush.

Several miles to the west and at the base of the continental slope, the next hole encountered the predicted sabkhas at the edge of the salt lake with their diagnostic anhydrite textures. Then followed in proper succession the lagoon with its alabaster and algal mats. But the cores cut from beneath the abyssal plain, and in the very last hole drilled, were the most breathtaking of all, for they contained a deposit that no one, even in a dream, had expected to survive the trip from the seabed to the ship without dissolving.

And from a distance it looked as if the roughnecks, to tease the anxious scientists, had delivered the dream in an empty translucent core liner. Close up and in better light, however, one was able to discern the outlines of long, thin cylinders of rock that were themselves almost completely transparent. Impatient as usual, Hsü reached into the tube and pulled out a piece that looked to Ryan exactly like an icicle. Hsü put it to his lips and licked it, then passed it around for all to enjoy. The taste of salt was unmistakable.

In technical parlance the salt would be later described in the initial reports of the expedition as the mineral halite with a composition of sodium chloride and containing zones enriched in magnesium and potassium. Miraculously, these highly soluble minerals endured the ten-thousand-foot elevator trip through the water because the tool pusher, at his own initiative, had hauled Core 134–10 to the rig floor with the accelerator pressed to the floor.

Later that night as the *Glomar Challenger* steamed to Lisbon, the exhausted scientists and drilling crew slept. They had been hard at work for fifty-five grueling twenty-four-hour days. Ryan, fearing that he had been deceived in a dream, returned to the core lab. On the table was the cylinder of salt that Hsü first licked. With a jeweler's saw Ryan delicately cut it in two to expose the hidden layering. Clearly visible on the fresh smooth face of rock was the cross-section of a desiccation crack, filled with salt crystals. Indeed, for one brief moment even the briny lake in the center of the Mediterranean had withered to an empty puddle!

FOR Ryan the implications of the expedition's results did not fully register until the shipboard reports were finished, the press conference was over, and he was back at the Lamont-Doherty Geological Observatory and in

*Bill Ryan inspects a slice of salt recovered by drilling into the floor
of a once-dried Mediterranean Sea*

contact with Cita and Hsü through the mail. He had just received a letter from a Russian scientist named I. S. Chumakov who had read about their results in an account in *The New York Times* that was excerpted in *Pravda*. Chumakov had been a member of a Soviet engineering team that had built an enormous dam across the Nile River at Aswan in Egypt. Chumakov had been in charge of drilling a series of bore holes into the Nubian bedrock from bank to bank in order to locate a secure foundation for the dam. When it came time to drill at the river's center line, the hole had penetrated the usual twenty to thirty feet of riverbed silt and sand but then continued another nine hundred feet in these sediments before the bit struck the granite substrate. The engineers had discovered an extraordinarily deep and narrow gorge belonging to an ancient hidden river. More astonishing was the recovery of deep-sea ooze in the bottom of the gorge, sandwiched between Nile mud and granite bedrock. The ooze was exactly the same age as the sediment cored by the *Glomar Challenger* right above the anhydrite.

Chumakov realized when he read the report in *Pravda* that this ancient river beneath the Nile was in fact a spaghetti-thin arm of the Mediterranean as it existed about five million years ago. What was perplexing was the location of the gorge more than six hundred miles inland from the present coast. However, there had been no doubt about this deep canyon connecting with the Mediterranean. In the ooze he had seen not only the tiny shells of marine plankton but shark's teeth! In order to reconcile how salt water could have invaded a stream so far inland, Chumakov constructed an explanation just as outlandish as the evidence in his cores. He concluded on his own that the surface of the Mediterranean had once dropped more than five thousand feet below its present level.

While the Mediterranean was drying up, the Nile was incising a deep valley to continually adjust its stream gradient to the depressed coastline. When flooding eventually refilled the Mediterranean and restored the surface of the sea to its former level, the gorge drowned, causing it to become a marine estuary. So swiftly had the salt water threaded its way into the interior of Africa that the Nile, with its own infilling from annual floods, could not keep pace to prevent the Mediterranean's invasion of Aswan. Chumakov's letter was greeted by the *Glomar Challenger* scientists as a stunning confirmation of a deep sea transformed into a desert and then back to a sea again.

Some months later the Phillips Petroleum Company invited Ryan to Bartlesville, Oklahoma, to present the highlights of the expedition to their exploration geologists. It was to be a brief visit. After the talk in which Ryan presented a sketch of the Mediterranean desert landscape and indicated how the desiccation was supported by the deep drilling, he was asked to delay his return. The following morning Ryan's host picked him up at his hotel and escorted him through tight security into the research complex where new oil prospects were evaluated in secret. Phillips, in partnership with Italian and Egyptian oil companies, had drilled the subsurface of the Nile delta without success. In the light of the *Glomar Challenger*'s results, the explanation was straightforward. In the maze of reflection profiles laid out before Ryan, a vast landscape of buried river valleys appeared directly under Alexandria on the delta and extended inland well beyond the pyramid of Cheops at Giza on the outskirts of Cairo. In these ancient badlands the trunk of the ancient Nile riverbed was discernible along with dozens of its major tributaries. Apparently during the drawdown of the Mediterranean through evaporation, the entire continental margin of North Africa had been exposed as land and had become severely eroded. No wonder there was no oil and gas. All the strata that might have been reservoirs or traps for hydrocarbons had been washed away by streams.

More and more supporting evidence poured in. Bone-hunting paleontologists from the American Museum of Natural History in New York discovered some of our very distant African primate ancestors in southern Spain. They had come over from Africa, presumably across the barrier that had cut off the Atlantic Ocean from the Mediterranean, and had allowed the latter to dry out. On the island of Cyprus, investigators from University College in London excavated the skeletons of elephants and hippopotamuses from graveyards 5.5 million years old. These mammals were not the usual multiton behemoths of East Africa. They were pygmies that you could have picked up and carried around in your arms as pets. Apparently they had wandered down a distributary channel of the Nile and deep into the empty desert basin to inhabit lakeside swamps and neighboring savanna. In the novel ecological setting on the floor of the broiling hot eastern Mediterranean, the elephants and hippopotamuses had evolved through natural selection to a dwarf form that could cope with the hellish conditions. Their skeletons had been fossilized in the deposits of the riverbeds. Later the ongoing collision of the African and Asian continents had uplifted the buried northern rim of a lake, long turned into sedimentary rock, and thrust it into the landscape that would one day become the Pentadaktylos mountain range of northern Cyprus.

ONE Sunday afternoon in 1972 an amateur fossil collector dug into a hillside outcrop of gypsum-bearing rock in the Tarano Valley in the Piedmont region of northern Italy. He peered at the inside face of the thinly laminated anhydrite rock that had just split apart with the blow of his hammer and saw a specimen of an ancient eel, the outlines of its entire body and fins splendidly preserved. The fossilization in this rock was exceptional because the environment at the time the sediment was laid down had been a briny lagoon whose tranquil bottom waters were devoid of oxygen. No scavengers had been able to tolerate such conditions.

When the quarried slab was delivered to Carlo Sturani, an articulate and energetic professor of paleontology at the Institute of Geology of the University of Turin, he knew immediately that it was equivalent in age to the Gessoso Solfifera of Sicily and the anhydrite and salt recently discovered by the *Glomar Challenger*. He visited the cliff to undertake a detailed investigation of a succession of fossil-rich rocks. Along with more eels he found foraminifera, corals, echinoderms, conch, herring, small flounder, dragonflies, leaves, acorns, land turtles, freshwater reeds, and roots of trees still in place. In a three-hundred-foot cliff Sturani could observe a moderately deep former sea that had dried out and become a tidal flat with algae and mud cracks. Then it became a shallow lagoon so concentrated by evaporation that its brine precipitated massive banks of selenite from which the first eel had been discovered. After a while the lagoon turned into a brackish lake, sometimes filled with freshwater. Then the lake withered into a peat bog as the region progressed from marshland to a sequoia forest. Abruptly, in the span of a tenth of an inch of rock, it was once again an open deep sea situated far from land. The transformation from sea to land and back to sea had taken less than half a million years. Except for those privileged to have been on the *Glomar Challenger,* no one else had ever expected that a major sea such as the Mediterranean could have evaporated so rapidly and refilled so quickly.

Intrigued that this sequence, from a distal margin of the ancient Mediterranean, mimicked the one recovered by seabed drilling, Sturani brought his results to a conference in the Netherlands where he hoped to meet either Hsü, Cita, or Ryan. As it turned out, all three had been invited to the same meeting to present to the European community the initial reports of their expedition.

Sturani projected a slide of his eel onto the giant screen of the conference hall and described its appearance.

The backbone of this young eel was wrinkled in an unusual zig-zag pattern. This is a post mortal feature, due to a shrinkage of the soft tissues: the rigid backbone was forced to bend in order to cope with the shrinkage, which reaches as much as a fourth of the original length. In turn the shrinkage is a result of the dehydration.

As Ryan listened, Sturani elaborated his interpretation that the eel had entered into a lagoon filled with brine except for a thin cap of less dense freshwater. It had perished there, settled onto the oxygen-depleted sterile bottom, and been pickled. The brine had sucked all the water from the tissues inside the eel through its skin via the well-known osmotic process, thus causing the body to shrink and the backbone to compress into the contorted skeleton so vivid on the screen.

Ryan's thoughts raced back to his expedition to the Mediterranean aboard the *Chain* a dozen years earlier when he had stood watch during an echo-sounding survey of the Strait of Gibraltar. He remembered vividly that at the entrance to the Mediterranean Sea there had been hundreds of small, color-ful Moroccan boats fishing for eels. It had been explained to him at the time that this was the eels' breeding ground. Now Ryan realized why. The dam, which five million years ago had kept the Atlantic water out of the Mediterranean, had been located at Gibraltar. It had also been a barrier to the eels. The high wall had effectively prevented the Mediterranean eels from joining their Atlantic relatives.

It is well known that eels living today in European rivers that flow into the Baltic Sea and the Atlantic Ocean migrate to a breeding ground in the Sargasso Sea south of Bermuda, whereas those in European and North African rivers that flow into the Mediterranean reproduce east of Gibraltar. Their copulation takes place at the foot of the long-destroyed dam in the Gibraltar Strait. Ryan wondered if the memory of their former isolation had been stored genetically for millions of years.

Dick Benson, a specialist in tiny aquatic crustaceans at the National Museum of Natural History at the Smithsonian Institution in Washington, D.C., had his turn at the podium on the final day of the conference. For him "the evidence for shallow seas or sea-lakes of rapidly changing salinity during the Messinian [a stratigraphic name for the time interval between 7.2 and 5.4 million years ago during which the Mediterranean became separated from the Atlantic] appeared very convincing." In fact, the news was a bit anticli-mactic. For Benson the greatest excitement was not from the anhydrite and desert sands but from the microscopic creatures in the marine oozes below and above the salt. In his analyses of the *Glomar Challenger* cores, he had found tiny aquatic crustaceans belonging to species without eyesight that generally live in the cold abyssal waters of the deep ocean. These he had discovered in the marine sediments, which like two pieces of bread form a sandwich bracketing the time between the initial evaporation of the Mediterranean and its reflooding. These faunas could have returned to the Mediterranean within a few thousand years after the floodgate was opened in only one way.

The Gibraltar dam must have collapsed catastrophically. Salt water from the "bathyal realm" of the Atlantic had inundated the Mediterranean desert at the pace of thousands of Niagara Falls. In the process the raging torrent had eroded the former barrier, incising the breach to perhaps one thousand feet below the level of the inrushing Atlantic. Such a deep opening was considered necessary to siphon in the blind crustaceans from the abyss of the North Atlantic and deliver them, enveloped in salt water no warmer than 40 degrees Fahrenheit, to the rapidly filling basins of the Mediterranean. Had the portal not been wide open, Benson argued, an entirely different set of tiny crustaceans would have been found in the *Glomar Challenger* cores.

Although no humans lived five million years ago, had any been present, they would have witnessed the Mediterranean desert disappearing permanently beneath a mile of salt water in a matter of a single human lifetime.

CHAPTER EIGHT

Vanished Deserts

ON a bright fall afternoon in 1971, John Dewey, a ruddy-faced, hyperactive Englishman with a fresh Ph.D. from University College in London, picked up a kidney-shaped paper cutout representing a region composed of billion-year-old rocks that was once part of the belly of Europe. He slid the piece across the tectonic map of the Alps he was constructing on his drafting table and joined it to a sliver of Africa now residing in the Rhodope Mountains of Bulgaria. In so doing he would unwittingly set in motion a search for another catastrophic flood, not so ancient as the one that filled the Mediterranean five million years ago but in fact so recent that it was probably witnessed by humans.

Dewey had left home and crossed the Atlantic five years earlier to find out more about sea floor spreading. At Columbia University he was greeted by Marshall Kay, an illustrious professor nearing retirement who was about to see his life's work in geological theory replaced by the new paradigm. The newly discovered spreading of the ocean floor had revealed that the outer skin of the Earth did not remain rooted in place but instead slid across our planet's viscous interior like ice on a river. Such relatively large-scale movements, occurring at speeds measured in inches per year, propelled the continents and caused them occasionally to slam into each other, pushing up the mountain chains as they collided.

Marshall Kay reveled in the excitement, proud that a Columbia graduate student had just published an attention-gathering paper on ocean floor magnetic anomalies in the prestigious journal *Science*. These anomalies corresponded to individual bands of alternating normal and reversed magnetism in the ocean crust. When plotted on a map, they had revealed a striking pattern of linear and symmetrical stripes running along the floor of the

Pacific Ocean midway between New Zealand and Antarctica. The pattern in the Pacific turned out to be identical to one just discovered in the center of the north Atlantic Ocean south of Iceland by an aeromagnetic survey of the U.S. Navy. Such a consistent symmetry in both oceans perfectly fit the bold idea that while the continents separated, new crust was formed along a permanent crack that comprised the separation. There molten lava erupted, cooled into rock, and formed the volcanic substrate that preserved both the strength and direction of the earth's magnetic field. Following the announcement in *Science* there was a flurry of activity to search the global ocean for the same pattern and to extend the mapping of the magnetic stripes worldwide.

Kay introduced his young British guest to Walter Pitman, the author of the *Science* paper. Pitman was a few months away from the completion of his Ph.D. thesis. He had entered oceanography six years earlier and had spent nearly a year and a half at sea, including a ten-month voyage on a three-masted schooner, the research vessel *Vema*. The ship was operated by the Lamont Geological Observatory, a premier research center and part of Columbia University where Kay taught and Pitman studied.

Kay hoped Dewey would see for himself the compelling evidence for continental drift provided by the patterns of magnetization in the ocean's crust. Kay was also eager to show off the modern facilities of the Lamont research center where more than one hundred investigators were working in all disciplines of the earth sciences, from probing the solid earth interior to the fluid ocean and atmosphere. This satellite campus of the university was also home base for about another hundred graduate students carrying out their research in the far corners of the world—from the bottom of ocean trenches to the poles, and even to the surface of the moon.

This voyage on the *Vema* took Pitman around the world. During the year prior to Dewey's visit, Pitman had undertaken a second research voyage on which he had conducted the magnetic survey across the Pacific Antarctic Ridge, which resulted in his graph of symmetric anomalies. Pitman showed his work to Dewey, explaining that if the area of the oceans were increasing, as the seafloor magnetic mapping demonstrated, then certainly the continents at the oceans' edges had to be moving with respect to each other.

Over the next two years Pitman completed and published a time-lapse reconstruction of the separation of Africa and Europe from North America during the past two hundred million years. Using the magnetic reversals and bathymetric mapping of the North Atlantic sea floor, he had determined the trajectories of the continents with remarkable precision. By then John Dewey was at Lamont on a yearlong leave from University College in London.

Ryan and Pitman had been colleagues at Lamont since 1962 when they invited Dewey to join them in a bold undertaking to unwrap the evolution of the Alps, a relatively young mountain belt that extends in an almost unbroken arc across the southern edge of Europe and Asia—starting from the Betic Cordillera of Spain; passing through France, Switzerland, and Austria, and around the Carpathians of the Czech Republic, Slovakia, and Hungary; down to the Hellenides of Yugoslavia and Greece, including the Balkans of Bulgaria; along the Taurus of Anatolia; and reaching through the Zagros Mountains of Iran where the ancient Behistun Rock of King Darius is located.

The three scientists decided to tackle anew the issue of building mountains with colliding continents and disappearing oceans, by using the relative motions of the drifting plates. File cabinets overflowed as they gathered manuscripts, treatises, field notes, structural maps, computer output, satellite photographs and stratigraphic columns. They met often in Dewey's lab for three- or four-day binges to pore over the data. Their goal was to document the closing up, in steps of twenty million years, of the formerly wide ancient sea known by earth historians as Tethys, the mythical wife of Oceanus, which had once stretched from the Caribbean to the Himalayas. Their reconstruction highlighted the collision of the African promontory of Morocco with the European promontory of Spain seven million years ago, the event that sealed the Atlantic from the Mediterranean and led to the transformation of the latter from a sea to a desert.

The three earth scientists were coping with small slivers ripped from one continent and attached to the other. One bothersome region was the Black Sea basin; its relatively thin crust, compared to the adjacent underbelly of Ukraine, suggested that this depression was a remnant of an old ocean that had somehow escaped the collision and was now trapped behind the rising Pontide Mountains of Turkey. There it sat surrounded by continent on all sides with only a long, narrow bottleneck connecting it to the Mediterranean. This bottleneck reminded Ryan of the Strait of Gibraltar. He wondered whether the Black Sea might once have dried up just like its big cousin, the Mediterranean. Not yet finished with his reports from the *Glomar Challenger* expedition, which had concluded only recently, Ryan felt disloyal to his former shipmates for his present absorption in the tectonics of mountain building instead of anhydrite and ooze. He commented on his distraction to the others.

Somewhat sarcastically, Dewey challenged, "Bill, do you suppose the catastrophic flood that filled the Mediterranean might have been the one Noah escaped in his ark?" Pitman chimed in, "You know, I bet there were other floods like the Mediterranean flood that were not caused by rain but by the permanent filling of a large enclosed depression."

Dewey considered this a moment, then challenged again: "Well, if you think so, give me a plausible candidate for Noah's flood." Since they had been concentrating on continental collisions that had occurred twenty million years ago, Dewey cautioned, "Now don't make Noah a prehistoric ape man. Offer me something recent and not ridiculous."

The Mediterranean flood was not a candidate. If the flood described in Genesis and *Gilgamesh* had been a real historic event, it would have to have been witnessed by modern humans—people with spoken language, the ability to plan ahead, an advanced cultural toolkit, but, most important, with a lifestyle endangered by the consequences of a flood. *Homo sapiens sapiens* apparently emerged from Africa roughly one hundred thousand years ago and began to experiment with settled villages and the cultivation of cereals in the Fertile Crescent of Mesopotamia only at the end of the last ice age, around twelve thousand years ago. So although the flooding of the Mediterranean from a Gibraltar waterfall was too ancient an event, it could still serve as a useful role model by suggesting which phenomena one would look for in the geological record for a much younger basin-filling catastrophe.

T H E drying out of the Mediterranean proved that such saltwater floods "were rare, but real events." The question for Pitman and Ryan was not whether there had been other such floods but where one would go to find an appropriate setting. The most obvious answer was the submerged floors of modern seas with narrow connections to other seas. One consequence of a drawdown of an enclosed sea or lake for later flooding is the one-way flow of streams and ocean spillways into the depression. Creatures following the trail of moving water will be guided downstream and eventually to the shores of lakes and brine pools. Some completely enclosed lakes might indeed have been fresh and potable, as attested to by mollusks in the eastern Mediterranean and by the types of fish and reeds found by Carlo Sturani in the Piedmont region of Italy. These lakes and their tributary streams would be oases in an otherwise inhospitable world. Like the pygmy elephants and dwarf snails, the actors on this stage should not be hard to distinguish from a normal crowd.

Lately Ryan had been thinking about a more comprehensive set of predictions to test the drying out of the deep Mediterranean. He was ready to try some of his ideas on Dewey. Ryan admired Dewey's wealth of experience and critical attitude. Furthermore, Dewey was someone who had no personal ax to grind for or against the outlandish desiccation hypothesis.

"First you would look for an arid setting," Ryan replied. He knew that deserts only developed under specific conditions. Even parts of Antarctica, bathed in frozen water, have remained bone dry and without new snow for centuries. If one were searching for drowned deserts, then one must pay close attention to climate. Luckily, the clue to an arid setting can be found outside the immediate dust bowl. The Ice Age expansion of deserts all across Asia could be detected in products carried by the wind all the way to the Greenland ice cap. The same glacial wind blew pollen from plants and trees far out into the ocean where it settled in a deep-sea sedimentary repository. Sagebrush in the absence of oak and pine was a traditional indicator of a dry region. Ryan said to Dewey, "With a little bit of detective work one can always find a finger print or two of a vanished desert."

Recalling I. S. Chumakov's hidden Nile gorge, Ryan also proposed another approach: "One might also look for signs of entrenched river valleys." Generally a river flows across the surface of its floodplain and builds a delta. Many of the ancient seaports of Mesopotamia constructed in the third and second millennia B.C. at the sea's edge, such as Uruk, Ur, and Eridu, are now stranded eighty or more miles inland by the growth of younger deltas of the Tigris and Euphrates rivers. Over time a youthful river becomes stacked on the buried channels and levees of its ancestors. When either the land rises or the sea falls, however, the riverbed cuts down into its foundations. The new, steeper trajectory causes the stream to flow faster and with more energy, causing channel incision into the floodplain. A desiccating sea will be fed by rivers flowing through deep valleys. The Grand Canyon in northern Arizona is an extreme example of such erosion. In this case the Colorado Plateau was uplifted thousands of feet above the Pacific Ocean some ten million years ago. Beneath Cairo, the ancestral Nile flowed six million years ago through a chasm just as impressive as the Grand Canyon!

Prodded further, Ryan came up with several methods he would use to recognize deeply incised rivers. One was to look for massive amounts of the eroded material transported by currents to the basin center. As each river cut down into its stream bed it would exhume the fill of older streams. If the entrenchment was severe, the gravel and coarse sand of the previously buried channels would be transported to the bottom of the sea along with shells of freshwater snails and much plant or woody debris. At the last drill site west of Sardinia, the *Glomar Challenger* had cored pebbles delivered by flash floods in sudden desert thundershowers.

Other clues might be better seen in aerial photographs or topographic maps. The Danube River in southeast Europe displays at least seven levels of terraces (abandoned river floors). The higher terraces are of successively older ages. Each younger stream bed evidently remained nested into the floor of its parent stream. Such continuous confinement suggests that as the river cut down into its ancestor channels, it forever lacked the capacity to breach its banks and escape to another route across the alluvial plain. Such reduction in flooding capacity would be predicted in an environment becoming more and more arid through time.

Dewey pressed his colleagues hard to acquire even more subtle evidence. He prodded, "Okay, you find a desert and deeply incised rivers, and then what else do you look for?" This was a topic that Ryan had discussed with Maria Cita on many occasions, both on the drill ship and in correspondence.

There would be a substantial replacement of the fauna. Conditions in the ocean are actually rather unstable. Marine plankton are not especially tolerant in their adaptability to variations in temperature and salinity. However, near-shore fauna are often much hardier in the face of environmental change. In technical jargon, the fauna and flora able to survive for a long time under varying conditions are "cosmopolitan."

Ryan elaborated that before a sea started to dry up, one would find a diverse population of specialized creatures, each individually adapted to its own environmental niche and most unable to survive elsewhere. In the desert phase they would be replaced by more opportunistic adventurers and the "cosmopolitan" generalists. He and Cita had reviewed many case studies of the past geological record, which showed that, as the environment skewed toward the extreme, there would eventually be only one or two holdouts before total sterility.

As an example, blue-green algae were among the earliest life forms to evolve on Earth. For a period of a billion or more years they and microscopic bacteria dominated the planet. The algae manufactured the oxygen of our atmosphere. When other life-forms began to evolve to greater complexity in a world made hospitable by the algae, the first predators appeared. The algae, which had made the predators' evolution possible, practically vanished from the face of the Earth as their victims. The algae have managed to reemerge only rarely and fleetingly. One of the last times was when they colonized the shorelines of Mediterranean brine pools where the salt content of the drying sea had reached such proportions that all the grazers of algae had been locally excluded. Comparable colonies of algae exist today in one tiny spot at Shark's Bay, Australia. There, in the interior of an isolated lagoon cut off from the Indian Ocean, the seawater salinity is ten times its normal value. In trying to swim there, as in the Dead Sea, one bobs helplessly like a cork set adrift in a bathtub.

In fact, as soon as the Mediterranean had been shut off from the Atlantic, the hundreds of specialized species populating its vast colorful coral reefs were rapidly annihilated. Only one, *Porites,* was able to hang on as a reef builder—a feat achieved with tenacity and only for a relatively short time. Although perhaps not intuitive, stability is indeed the handmaiden of diversity in the long march through the corridors of planetary evolution. Global change provides the punctuation for the Earth's story.

At this point Pitman proposed that they consider an episode of aridity in the relatively recent past. "How about sometime since the melting of the last continental ice sheets?" he asked. Then without waiting for the reactions of the others, he volunteered: "I bet you would see caves and campsites abandoned as hunters and gatherers were induced to migrate elsewhere for survival."

Pitman was well informed about the Ice Age caves of France and Spain with their glorious wall paintings. He continued, "When excavating the interiors of an ancient tell, perhaps you would come across a gap in settlement sandwiched between overlying and underlying floors of occupation." Such a period of total sterility had occurred a number of times in the severe desert conditions that had occupied the Mediterranean for several hundred thousand years during its isolation from the Atlantic.

"All right," said Dewey. "You look for seas turned into deserts. You find deep river valleys and perhaps some weird inhabitants with strange behavior. But how do you find the flood itself?"

Ryan replied, "By the same sudden replacement of the fauna I just told you about!" The floodgate that had opened at Gibraltar five million years ago allowed the entry of marine plankton, fish, and sea mammals to replace completely the desert vagabonds and squatters. The filling of this huge depression may have taken less than a century. The creatures that inhabited the new seabed took a bit longer to reach their final destinations than those that could swim or float passively. But the replacement was essentially accomplished in a flash of geological time. Ryan added, "Of course, those species in the outside ocean that became extinct after the dam had been erected were no longer around to celebrate the good old times once the barrier came tumbling down."

"And what became of those in the way of the flood?" asked Dewey.

Ryan thought for a minute. He then responded, "Charles Lyell reported a whole bunch of mammals suddenly appearing out of nowhere on the Mediterranean islands, such as Sicily, Sardinia, and Malta. Maria Cita mentioned this to me on the *Glomar Challenger*. There are elephants and hippos in Cyprus and Crete. In scrambling to high ground in response to the flood they arrived in places they had never been before."

To the three scientists gathered around their languishing reconstruction of the Alps, the most useful clue for identifying Noah's Flood had emerged. It was to look for human refugees driven into exile by the water reclaiming the territory they had formerly occupied.

CHAPTER NINE

Pontus Axenus

BILL Ryan had been giving considerable thought to the subject of faunal replacement. Not only had he discussed it with Maria Cita on the *Glomar Challenger,* but he had also started to read about it in scientific journals. Somehow, however, he missed the very story he was hoping to find.

In early October 1970, *The New York Times* reported under a four-column headline that the Mediterranean Sea had been "a land-locked basin where, through evaporation, little was left but a salt plain." Ryan had been so absorbed in this announcement of the results of his drilling expedition that he failed to see an article published simultaneously in the journal *Science,* the prestigious weekly publication of the American Association for the Advancement of Science. It concerned the recent sedimentary history of the Black Sea. Beyond lacking a catchy title, any reader would have had to probe deep into its jargon-filled text to learn that something remarkable had been discovered. Yet there in both words and pictures lay a finding as significant as the great Mediterranean flood.

A replacement of fauna had indeed happened. And it had transpired not five million years ago but as recently as the last ice age when modern man roamed widely throughout both Europe and southwest Asia. The reported metamorphosis was not of a sea becoming a desert and then returning to a sea, however. It was instead of a former sea that had turned into the world's largest freshwater lake and then back to a sea. Its fish, plankton, crustaceans, mollusks, and coastal plants had all been affected. Those species that had formerly colonized a saltwater habitat were replaced species by species by those adapted to freshwater. Then the freshwater species had disappeared, and in their place the saltwater ones had returned.

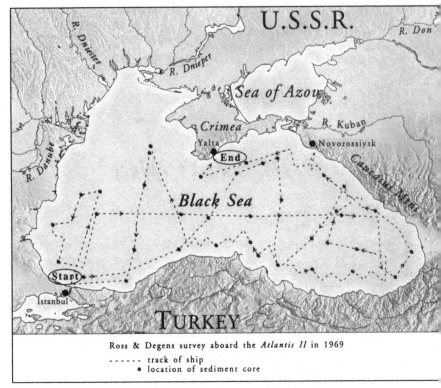

Ross & Degens survey aboard the *Atlantis II* in 1969
- - - - - - track of ship
• location of sediment core

The Atlantis II *mapping and sampling the seabed of the Black Sea in 1969*

The lake had persisted for thousands of years. It vanished once it reconnected with the Aegean arm of its neighboring Mediterranean Sea. The *Science* article noted that the reentry of salt water had occurred sometime between twelve and seven thousand years ago. Apparently not a single species of the earlier freshwater environment survived the transition back to a sea. Most of the lake inhabitants were annihilated as the salt content rose. Those that survived were able to hang on only by seeking refuge in the Black Sea rivers.

The increase in salinity created the Black Sea that we know today—the body of water explored during the mythical voyage of Jason and his Argonaut crew in their quest for the golden fleece. Until this era of the epic hero, the land surrounding the Black Sea had remained essentially *terra incognita*. Historical records date the foundation of the earliest seaports to the mid-eighth century B.C. These and other frontier trading posts established in the course of the following century gave the merchants from the Aegean and Ionian seas access to rich granaries, nearly endless stands of timber, and hoards of precious metals.

In his *Natural History,* Pliny the Elder described the Black Sea as "having swallowed up a large area of land which retreated before it." He gave it the name Pontus Axenus (from the Greek *axenos,* meaning inhospitable), mentioning that without abundant islands to shelter sailors in the notoriously brutal storms, an open crossing of such a vast body of water would almost certainly have brought peril. The Turks of the Middle Ages, in similar fright of being shipwrecked on its wave-tossed surface, called it Karadeniz, the black sea, harbinger of death.

The research announced in *Science* had developed quite by mischance. A team of geologists and chemists at the renowned Woods Hole Oceanographic Institution on Cape Cod, Massachusetts, had hoped to return to the Red Sea in the summer of 1967 to follow up some earlier discoveries. But the scientists' plans were canceled when war broke out between Egypt and Israel. So the researchers decided, after steaming into the eastern Mediterranean Sea, that instead of turning right to Port Said and the Suez Canal, they would go left into the Aegean and on to the Black Sea. The Woods Hole team was led by David Ross and Egon Degens. Their ship, *Atlantis II,* was commanded by Captain Edmund Hiller, the same skipper who had directed the *Chain* when it sailed into the Black Sea in 1961 with Bill Ryan aboard.

In place of a greeting from a Soviet escort destroyer after passage through the Bosporus, *Atlantis II* gained the attention of a Soviet four-engine bomber that leveled off at masthead height on the first of a dozen or more terrifying passes. When the roar of its turbine engines at last faded away, the sky darkened, and the wind began to whip the sea into a tempest. By nightfall a gale had intensified to near hurricane strength.

With characteristic impatience Ross and Degens had already stopped *Atlantis II* for a first attempt to sample the bottom sediments, even while still being buzzed from the sky. The coring tube had reached the seabed when Captain Hiller decided that the exercise should be aborted. However, with the ship heaving in the mounting swell, bringing the heavy gear aboard was fraught with serious risk. The forward thruster alone, unassisted by the action of water flowing past the ship's rudder, was slowly losing its ability to keep the bow of the ship pointed into the wind. Hiller was sure that if *Atlantis II* fell off her heading and slipped into the trough of the sea, they would then roll at the mercy of the storm.

Fortunately, after an arduous struggle, the seasoned deckhands delivered the coring device back on board and lashed the mud-filled tube to the deck.

By mid-morning the next day, when Degens strolled out on the afterdeck, the sky had cleared and the surrounding sea was subsiding. Wandering over to the coring device, immobile in wood chocks and chains, he noticed jet black mud oozing from the end of the tube despite a valve installed to prevent such an incident. A sizable puddle of the soft gelatinous material had accumulated on the deck. Disheartened, Degens checked whether the valve had properly closed. It was indeed shut, but the gunk was still leaking out; something was forcefully extruding it!

A sniff of the air solved the mystery. An invisible gas, a poisonous combination of hydrogen sulfide and methane, slowly vaporizing from a former icelike state in the abyssal graveyard of the sea, was expelling the core material. To disassemble the core tube safely in the open air Degens used an electric-powered screwdriver. He removed the twenty-foot metal sheath and scraped the exposed face of the recovered sediment with a clean spatula to expose its internal layering. The top forty inches was a dark black, jellylike mud. Below the material was a light gray clay. The black mud revealed hair-thin white bands at a density of approximately one hundred stripes per inch, stacked one upon the other.

Everywhere *Atlantis II* went to recover additional cores the sequence of black mud overlying light gray clay was the same. The black mud was gorged with plant and animal remains. At some levels its content of organic carbon reached an astonishing 50 percent of the bulk dry weight of the sediment, whereas in typical sediment from oceans like the Atlantic or Pacific the enrichment is only a tiny fraction of 1 percent. When viewed through a microscope at high magnification, the thin white bands turned out to be composed entirely of the skeletal framework of a single species of plankton that lived near the sea surface where it could be bathed in sunlight.

More powerful magnification of the organic soup revealed membranes of proteinlike structures. No one expected that the long-dead tissue and its amino acid building blocks could be preserved in such exquisite detail. The organic residues were thousands of times more enriched in the gelatinous black mud than in the underlying light gray clay.

At first glance the older clay deposit below the black mud had no visual treasures. When the clay was put in a press to squeeze out its pore fluids, however, the extracted water was fresh. In fact, it was so low in dissolved salts that had there been enough to drink, it might have tasted like mineral water from a mountain spring.

Ross and Degens composed a compelling story. In the past the Black Sea had transformed itself into a lake; it was linked to the Mediterranean only by a narrow outlet not deep enough to permit the continued entry of marine water from outside. This condition had been established more than twenty thousand years ago when ice sheets covered Scandinavia, northern Europe, and all of Canada. The building of continent-sized glaciers to a mile or more in thickness, as first envisioned by Louis Agassiz, had caused the withdrawal of global sea level below the level of the floor of the Dardanelles and Bosporus Straits. When the ice over Russia eventually began to thaw, much of its meltwater was delivered into the Black Sea's Ice Age lake. The light gray clay in Ross and Degens's cores came with the meltwater as a milky suspension via the Danube, Dnieper, Dniester, and Don rivers.

Eventually there came a time when the giant lake turned back into a saltwater sea. This event was marked in the *Atlantis II* sediment cores by the abrupt change from light gray clay to black mud and by the simultaneous disappearance of the skeletal remains of freshwater organisms and their replacement by creatures that lived only in brackish water or salt water. Ross and Degens had clearly detected a recent rapid faunal replacement. But these scientists had no reason to suspect any catastrophic inflow of the salt water. Their expedition was completed a full year ahead of the *Glomar Challenger*'s discovery of the Gibraltar waterfall that refilled the Mediterranean. The potential drying out of isolated inland seas was an idea that had not yet fermented.

Unbeknownst to Ross and Degens the Soviets had stumbled on a clue that the Black Sea's lake may have indeed partially dried out and shrunk to a size two-thirds its Ice Age extent near the end of the great thaw. In planning a railroad bridge across the strategic Kerch Strait, a four-mile-wide and twenty-mile-long passage into the Sea of Azov on the north coast of the Black Sea, Russian engineers had drilled a series of holes to learn the depth of the bedrock. They suspected the riverbed sediments were thin, as was discovered by the Soviet drilling beneath the Nile River at Aswan, directed by I. S. Chumakov. However, another chasm appeared that reached to more than two hundred feet beneath the strait's bottom. The deposits in the axis of this gorge were the sand, gravel, and snails diagnostic of a terrestrial streambed. There was no doubt about it. The entire Sea of Azov to the north had once been a tract of dry land across which a river (probably the ancient Don River) had cut its valley for a distance of more than one hundred miles in order the reach the shoreline of the shrinking lake.

This vast landscape had then drowned. The Soviet scientists found in the soft mud accumulating on top of the streambed deposits tiny mollusks with both halves of their seashells still attached by their hinge. Had these specimens been washed out of older nearby strata as contaminants rather than having been living dwellers of a new and deep estuarian environment, their delicate valves would have been scattered long before. These creatures must have lived in a substantial depth of water so as to have been sheltered from the disturbance of storm waves and surface currents. In examining the pore water in the mud containing the mollusks, the researchers realized that not only had the passage from a flowing river to a deep estuary been rather sudden, but the water that had invaded the land was of a different composition—enriched sodium, chlorine, and magnesium. The mollusks themselves were a replacement fauna carried in from the Mediterranean. It did not take long for the Soviets to recognize that the sediment cores from the Black Sea's perimeter contained the same abrupt transition from fresh to salty, just as *Atlantis II* had found in the central abyss of the Black Sea.

Keeping track of the sequence of sediments in the boreholes, the inquisitive researchers from Moscow State University began to investigate a large number of sediment cores recovered from the now-submerged seabed south of the Kerch Strait and reaching westward around the Crimea along the Black Sea coasts of Ukraine, Romania, and Bulgaria. A similar picture appeared. Outward from the shore for ten to more than a hundred miles the cores brought up a dark-colored mud layer, three or so feet thick and overlying a former terrestrial landscape consisting of river channels, steppe grasslands, and desert sand. Along the outer edge of the continental shelf the university scientists soon discovered beach deposits at 350 feet below today's sea level.

The Soviets added some important embellishments to the history tale initiated by Ross and Degens. They could demonstrate that when the Black Sea was a lake, its shoreline had migrated far beyond the present coast. Rivers flowing into the falling surface of this body of water had deepened and lengthened their pathways to maintain a steady downstream gradient all the way to their deltas. In the process of traversing the emerging terrestrial landscape, the Don River, in particular, had incised the valley encountered by the astonished civil engineers beneath Kerch Strait.

The Soviet scientists could also demonstrate that the arrival of the replacement marine fauna and flora was accompanied by a submergence of the Black Sea margin. Although they were not able to tell exactly when and how fast this flooding had occurred, they described it as rapid and said it had produced a rather impressive covering of a former terrestrial landscape. The big unknown in the case of the Black Sea was whether the surface of the Ice Age lake had ever lain below its outlet. Such a condition—the necessary precursor for a catastrophic flood—could only have happened under a regional aridity of the type that contributed to the desiccation of the Mediterranean. In other words, the lake would have had to shrivel in response to a regional drying out. Unfortunately, the past climate of the Black Sea periphery was poorly known. Its prehistoric record, retained in the soils of the coastal nations, was just then being unraveled. Sharing those secrets with the West would be the next step in putting the full puzzle together.

CHAPTER TEN

Red Hill

Two men sharing an umbrella emerged from the Metro station. In
unison they carefully picked their way across the shiny asphalt surface of rue d'Assas, threading in and out of the traffic heavy again
after the summer holidays. The man in the clerical collar walked erect in
short strides, while his hunched-over companion, carrying a package that
must be kept dry, shuffled along in steadfast concentration, not daring to
lean out from under the protection of their canopy to enjoy the orchards
with beehives in the Luxembourg Garden or the smell of hearth-baked bread
wafting from the boulangerie they were just passing. Sightseeing in Paris
could wait for another day. The courier was anxious; his thoughts raced in
his head.

The pace of their footsteps picked up as the drizzle turned into a downpour. The umbrella gradually dissolved into a montage of others, which, like
gelatinous medusa in a choppy sea, floated down the boulevard leading to
the dark stone façade of the main hall of the law school. Mounting the steps
to its entrance, the two men seemed oblivious to a brightly colored banner
hanging from the cornice overhead and blazoned with an icon of prehistory.
Its lettering announced:

VIII INQUA

In the expansive foyer, delegates formed several queues for registration.
Many had arrived the day before to participate in the opening session. As
collegial greetings echoed from the travertine-paneled walls, the man in the
clerical collar closed the umbrella. He shook off its clinging drops of water
and cautioned his comrade, "It is better if I get the badge. I will tell those at

the desk that I have been invited to give the benediction. If anyone asks, you are carrying my prayer book."

Temporarily abandoned, the one clutching his parcel wandered over to a group of registrants gathering in front of a sign for last names starting with the letters A through D. He wondered which, if any, of the elderly gentlemen passing through this line might be the distinguished professor from Columbia University. While he was perusing the new arrivals, he was rejoined by his comrade who handed him a badge labeled JIŘÍ KUKLA. The conspirators then parted company, the impersonator in the priest's garb to return to the urban world, and the gate crasher to enter the prehistoric Ice Age.

Jiří Kukla was penniless but elated. The day before, a two-hundred-franc registration fee had been one more roadblock thwarting his odyssey. Now, thanks to a plot hatched by a fugitive from behind the Iron Curtain and a white lie, he was a registered delegate to the Eighth Congress of the International Association for Quaternary Research.

Nine years earlier, INQUA, as the group was known, had met in Warsaw for its sixth congress. At that time, on behalf of the Czechoslovak Academy of Sciences, Kukla had helped prepare a comprehensive synthesis of the Ice Age deposits of his homeland. He had carefully crafted the report so it would be of lasting value to both the geologist and the archaeologist. To his disappointment the assembly had been rather sparsely attended by Western scientists, especially the American oceanographers with whom Kukla desperately needed to exchange ideas.

At a second opportunity for interaction in 1968, with a fully convened International Geological Congress in its inaugural session in Prague, columns of Russian tanks had rolled through the streets to thwart Czechoslovakia's desire for democracy and capitalism. Since then, Kukla and his coworkers at the Geological Institute of the Academy had been painfully isolated. Their anguish was not simply over the loss of journals and correspondence from elsewhere in Europe and America but at their inability to communicate their own important findings.

Now an exit visa of sorts was tucked under Kukla's arm—a thick set of scientific manuscripts freshly printed in German and bound in a book so as not to arouse suspicion at the Czech border. Kukla had to find a receptive ear. Determined to fulfill his dream, he threaded his way through the corridors of the building in search of the symposium on Evolution of Shorelines.

A century and a quarter earlier Charles Maclaren, a Scotsman, had conceived a corollary to Louis Agassiz's revolutionary theory of a universal Ice Age. Maclaren had addressed a question arising out of Agassiz's assertion that a thick sheet of ice had covered at least two-sevenths of the globe. The unresolved issue at the time was the source of the water locked up in glacial ice; from where did it originate? Maclaren reasoned that it had to have come out of the ocean, first evaporated as moisture and later precipitated as snow. Using a thickness of one mile to submerge the summits of the Jura Alps in a blustery white carpet, he calculated that such ice sheet growth would be "an agent capable of producing a change of 350 feet on the level of the sea." Maclaren's brilliant intuition predicted that each advance and retreat of the continental ice sheets could be measured by the fall and rise of coastlines all around the world.

It was more than a hundred years later, however, that the possibility of using carbon 14 to obtain the age of organic material was initially realized by the Institute for Nuclear Studies at the University of Chicago. Now former shifting shorelines could be dated to fix the episodes of chill and thaw. Still, a major obstacle prevented the full exploitation of this new way of telling prehistoric time. Carbon 14 decayed at such a rapid rate that its clock effectively ran down after the passage of fifty thousand years—a span of time then thought to be a fraction of a single glacial cycle.

Kukla was sure that all the attendees of the Congress held to the classical European dogma of four major episodes of ice growth and decay during the past two million years. But his investigations of the composition of wind-blown soils fossilized in the walls of a brick quarry near his home as well as an entirely new way of timekeeping by using the ability of soils to retain a memory of the past reversals of the earth's magnetic field made him certain that a dozen full cycles had taken place in the past half-million years alone. Somewhere among the attendees was an American who was known to share the same idea of a faster and rhythmic pulsation to climate change but who explored the cycles from the sea.

The morning passed quickly. Most of the papers concerned the melting of the most recent glacial episode and the tracking of the rising sea over the past fifteen thousand years. The speakers came from Australia, Holland, England, France, Italy, and Scandinavia. They reviewed a number of ancient shorelines exposed along the rim of the Mediterranean and North seas.

When Kukla saw a young speaker go to the podium for the next talk, he decided to take a break. He returned to the room just as the delegates were leaving. In the front of the room he saw the same young man engaged in a heated argument with others, all of them being coaxed out the doorway. As they passed within earshot, Kukla overheard part of the debate: "Wally, I just can't accept that the hundred-and-twenty-five-thousand-year age of your reef is accurate enough to be evidence of Milankovitch cycles!"

Kukla froze in his tracks. His ear had registered the name Milankovitch, a famous Yugoslavian mathematician renowned for an astronomical theory of the Ice Age. Kukla strained to eavesdrop further. The cornered protagonist held his ground until the next objection, apologized, and then bolted. Kukla caught a fleeting glance at the name on his badge. This "pip-squeak" was the scientist he had traveled from Prague to meet.

For the remainder of the day Kukla searched everywhere. As the lights came on at the end of the afternoon, Kukla was thoroughly discouraged. He slouched down the curving steps to the ground floor. Reaching the exit, he followed the crowd into the street. Barely twenty feet ahead was the face etched into his memory, once again hemmed in by others in lively conversation. Kukla sprinted to catch up, and grabbed the startled American by the sleeve. "We must talk," Kukla said in English, one of the many languages that he spoke fluently. Although initially flustered, Wally Broecker responded with an inviting gesture when he saw that the inquisitor had come from behind the Iron Curtain. "Why don't you tag along? My friends and I are going back to our hotel and then out for dinner."

That evening was one that Kukla would never forget. He began his story in the Red Hill brickyard on the outskirts of Brno, the capital of Moravia. Ten years earlier the yard workers had stumbled on the bones of woolly mammoths adapted to the Ice Age climate that had descended upon most of Europe. The animals had been buried in limy soils deposited on the easterly slope of a terrace alongside a tributary of the Danube River. The quarrying equipment had exposed the soils in a vertical cross-section revealing a rhythmically bedded profile that consisted of alternating layers of dark reddish brown organic humus and light yellow windblown dust called "loess."

Kukla had been summoned because he was studying the same loess that had drifted into caves occupied by Stone Age hunters twenty to forty thousand years ago. The stunning 160-foot-thick sequence of multiple soil cycles in the west wall of the Red Hill quarry offered a rare opportunity to examine a continuous record of the Ice Age deposits. Each climate cycle from warm to cold was expressed as a sequence of gradational soil types reflecting the change from a moist, deciduous forest to an arid, frozen tundra cracked by a deeply penetrating permafrost. Midway through each cycle Kukla noted a level with numerous bands of fine windblown dust delivered in monstrous sandstorms of continental scale. He speculated that this dust fog could have shielded the ground from the sun's heat for weeks or even months at a time and might have had a refrigerating effect on Europe's climate. In the colder part of each cycle, the environment had become so dry that even the mightiest rivers had ceased flowing. The former streambeds had become choked with large dunes of windswept sand.

The INQUA congress in Paris was a not-to-be missed opportunity to inform the world about three astounding facts gleaned from his research on the soil profiles. First, the ice sheets that had repeatedly advanced southward from Scandinavia and across Europe were invariably accompanied by the growth of vast but temporary deserts throughout Russia and Ukraine, even extending into southeast Europe and the shores of the Black Sea. Second, the passage back from freezing cold to warm was abrupt in every cycle, perhaps lasting only centuries. In the brickyard wall the flip-flop occurred in a layer of soil no thicker than the width of his hand. The gradations always progressed in steps from warm to cold. Once the climate had reached its coldest extreme, however, it jumped back to warm, skipping all the intermediate steps. The climate cycles of gradual cooling and sudden warming, when plotted as wiggles on a graph, looked like the teeth of a saw. The third surprise was that the last cold snap was only ten thousand years ago, well after most of the last ice sheet had withered away by melting. Yet this brief return to freezing conditions had been perhaps the driest of them all.

Intrigued by the reference to slow cooling and rapid warming, Broecker interrupted his new acquaintance: "Well, if you were at my talk this morning, you would have seen the same saw-toothed pattern in the deep-sea sediments and the same periodicity." He then told Kukla the new findings of the American oceanographers.

Ships from Broecker's lab had been plying the oceans since the 1950s. It had been the custom to take one sediment core from the ocean floor every day at sea. By the time Kukla had set up his camp in the Brno brickyard, more than two thousand such cores from the depths of the Atlantic, Mediterranean, Caribbean, and Gulf of Mexico were interred at the Lamont-Doherty Geological Observatory in Palisades, New York. They resided in stacked metal trays in five stalls of a carriage house on the former estate of Thomas Lamont, the head of J. P. Morgan's banking empire. Lamont's widow had bequeathed their estate to Columbia University in 1949, and a short time later it housed the Lamont Geological Observatory.

Broecker told Kukla that just a few years back two Columbia graduate students had located magnetic reversals in the deep-sea cores. He started to explain that at times in the past the earth's magnetic field had flipped its polarity. A compass's needle that had once pointed north would have then pointed south. As Broecker began to elaborate how these reversal boundaries could be used to precisely date the climate record, Kukla interrupted and pulled out his recently bound book and opened to a center-seam fold-out. The others at the table were unable to conceal their amazement that the Czechs had also been using magnetic reversals for dating Ice Age climates. The Red Hill brickyard soils contained not only the epoch of normal polarity that sat in the upper part of the deep-sea cores but the underlying epoch of reversed polarity that only a handful of deep-sea cores had penetrated. Kukla's soils spanned the whole interval. To Broecker and his companions no other terrestrial section had ever revealed such a record—and this one apparently reached back 1.5 million years.

Kukla flipped to a color sketch of the uppermost soil cycles in the Brno quarry. With his dinner companions now riveted in anticipation, Kukla showed that just twenty thousand years ago when fully modern man, *Homo sapiens sapiens,* was painting scenes of the animal hunt on the ceilings of caves in southern France and Spain, the ice cap had extended southward from Scandinavia almost to the Alps just as Louis Agassiz had suspected. One clan of Ice Age hunters in Moravia had constructed dwellings near the conjunction of two rivers in the drainage basin of the Black Sea. Their homes, comprising the Upper Paleolithic settlement called Dolni Vestonice, had been built over pits chiseled from the frozen peat with hand axes. The roof of each tentlike structure was arched with poles, which were draped with animal skins stitched together with sinew. The hides were anchored to the ground by massive animal hipbones and by the skulls of reindeer, mammoths, and the occasional rhinoceros. Tusks provided fuel for hearths. The tool culture was exceptional for its fine leaf-shaped blades struck from a core of flint and for small bone carvings of animal heads, notably the wolf and the bear.

Kukla showed Broecker his sketches of the campsites reconstructed from the excavations and the artifacts left behind; he pointed out that even under the harshest glacial conditions there had always been an ice-free corridor for animal and human migrations between eastern and western Europe along the Danube valley. It was within this refuge that his colleagues had found the famous Paleolithic-age "Venus figurines" of amply proportioned women. To their complete amazement the Czech archaeologists also encountered unmistakable evidence of a brief interval of baking clay in fire to make pottery, a technology that would not reappear until well after the close of the Ice Age and then in locations far away in Anatolia and the Levant.

It was now evident to Broecker and the others that continental soils exposed in Kukla's Red Hill quarry contained a remarkably complete and detailed record of climate change that fully complemented the story deciphered from the deep-sea sediments and beach terraces. Broecker and his companions had learned that Ice Age climates were extremely dry and that vast desert and steppe landscape had embraced Eurasia south of its massive ice sheets. Such desert conditions across the watersheds of the Dniester, Dnieper, and Don rivers might have produced a loss of water through evaporation sufficient in magnitude to drop the level of the Black Sea freshwater lake below its outlet, thus explaining why the ancient shorelines found by the scientists of the Soviet Union had been so far from the present coast settings, which are today much deeper than the floor of the Bosporus portal. Yet at this point the idea that a Black Sea lake's surface remained below the surface of the external ocean until the prehistoric lake turned back into a sea had not even been hatched. For all anybody knew, the connection at the Bosporus had been the gentle affair envisioned by David Ross and Egon Degens. Still to be determined was whether there had been sustained aridity, without which there would be no depression for salt water to plummet into and therefore no catastrophic flood.

At the time of the Paris meeting the Mediterranean had not yet been drilled, and John Dewey had not sent Bill Ryan and Walter Pitman off on a hunt for Noah's Flood. In fact, two years would pass before Ryan and Pitman would discover Ross and Degens's paper in *Science* reporting on the Black Sea faunal turnover, and another few months would pass before they learned that there had been a rendezvous between Kukla and Broecker in Paris.

Yet eavesdropping on the Ice Age tales told at that dinner would have been both encouragement and good entertainment. Desert climates and global sea-level changes were the two key subjects laid out that evening as food for thought. They not only cast light on sudden faunal replacements and deeply carved riverbeds but also on the conditions necessary for a flood of the type that would have expelled people permanently from a homeland and thus created the impression that their world had been repopulated by its survivors. For example, the Mediterranean would never have dried out when its gate to the Atlantic was shut by the collision of Morocco with Spain were it not for climatic aridity. An inland sea in which evaporation did not exceed the supply of incoming water from rivers and rain would not dry up. Even taking evaporation into account, a dam unbreached by a rising exterior sea would keep any potential deluge at bay.

Jiří Kukla had unknowingly seeded the trail of Dewey's future inquiry with critical but not compelling evidence. His dust-filled caves and brick quarry had unveiled a past climate in southeast Europe capable of at least partially evaporating a landlocked sea. The aridity was manifest in giant sand dunes sculpted by wind in southern Moravia and eastern Slovakia. As a schoolboy, Kukla had raced his classmates up the steep lee slopes of the high mounds to gain a panoramic view of what was described by his teachers as a dried-out lake floor. Using the horns of the crescent-shaped dunes as paleo-windsocks, Kukla had conjured the image of sand blowing from Europe to Asia in an unrelenting stream over millennia. He had asked, "Where had it all come from? When did it stop?" Only later as a university student did he learn that the sandstorms abated after the arrival of the first Stone Age farmers. Peasants would still occasionally find knife blades and arrowheads belonging to their nomad ancestors washed out of the dune interiors in rainsqualls.

On the other hand, Wally Broecker had been tracking the fall and rise of the global ocean. Although it had taken some clever gadgetry and several false starts, the Columbia University geochemist had finally been able to date the high-water marks. For his "tide gauge" he had selected Barbados. This slowly emerging island in the Caribbean is surrounded by a flourishing coral reef. Each time the balmy interglacial climate changed back to cold, the expanding ice cover on land had left the reef high and dry. Without its life-supporting film of seawater the coral bed around the island perished and then decayed into a terrace of bleached rock. Anyone flying into Bridgetown and gazing down at the tropical sea can view a set of exposed reefs wrapped around the perimeter of this island like bright concentric rings.

Following his instincts as a chemist, Broecker found out that the coral, when alive, digested minute quantities of uranium dissolved in seawater. This naturally radioactive substance became mineralized in the limy "skeleton," the white and pink framework rock seen everywhere by snorkelers and scuba divers. Over time the uranium 234 would decay to thorium 230 at a known rate, with both the diminishing ^{234}U parent atoms and the growing number of ^{230}Th daughter atoms staying sealed in the coral skeleton. By measuring the relative amounts of these two isotopes in samples collected from dead coral, Broecker had at his disposal a marvelous $^{230}Th/^{234}U$ chronometer that told time much further back into prehistory than the radiogenic clock powered by carbon 14.

Broecker applied his timepiece to the Barbados reef rings. The precision of his homemade mass spectrometer allowed him to demonstrate persuasively that every rise of global sea level in the Barbados coral reef terraces began with the melting of the large ice caps of North America, Europe, and Asia. Broecker's dating of sea level made it possible to envision the hypothetical circumstance when a rapid rise of the global ocean might occur while an inland sea was in a desert condition, a moment before the completed warming flipped the climate from dry to humid.

In 1971 an extraordinary set of diplomatic circumstances that had been set in motion by Wally Broecker landed Kukla in New York for a sabbatical visit to Columbia University. After settling into his new lab at Lamont-Doherty, Kukla presented the first of many entertaining seminars on the Ice Age of Europe. A throng of the observatory's scientists and students was in attendance, eager to hear him introduce his "mighty loess deposits," rumor of which had been circulating around the campus in the wake of the INQUA congress in Paris.

Bill Ryan was among the audience in Lamont Hall to see the slides of thirty-foot-high sand dunes that had marched across the floor of a once desiccated river valley north of Prague during the extreme aridity of the last glacial cycle. Today the Elbe River (called the Labe by the Czechs) wends its way through this fertile green corridor that only twenty thousand years ago had been frozen tundra, barren of trees and the home of migrating herds of the woolly mammoth. What struck Ryan most unexpectedly, however, was evidence presented and documented by Kukla that sand had been still blowing and shifting in seasonal storms as recently as eight or seven thousand years ago—a time when middle Stone Age (Mesolithic) bands of humans had passed through this landscape hunting for red deer, wild pigs, and mountain goats. That young date to the aridity was the clue Ryan and Walter Pitman needed for a young flood affecting people already settled in villages and able to escape by boat.

Ryan had at last read Ross and Degens's report on their Black Sea sediment cores. Attending Kukla's lecture he recalled the date of nine thousand years before the present that the two Woods Hole scientists had given to the Mediterranean's saltwater invasion into the Black Sea. Although Ross and Degens had never envisioned a partly evaporated depression for this water to enter into, Ryan realized that if western Czechoslovakia was still semiarid up until eight thousand years ago, then evaporation might certainly have drawn down the Black Sea lake below its outlet.

Ryan became stimulated by Kukla's new revelation of relatively young windblown sand and dust. But as his mind pondered the evidence, it was dampened by the fact that neither Ross nor Degens, whose expertise on the Black Sea was vastly superior to his own, had ever suggested a flood to explain what they saw in their cores of black mud. Neither had the Soviet scientists who had stumbled on the submerged shorelines and the deeply cut stream valleys in the area. In their judgment the same thaw of the great ice caps that had produced the meltwater to cause the rising global sea level and the drowning of Barbados reefs had also sent much of that meltwater down rivers to the Black Sea. The torrents streaming through the Don,

Dnieper, and Dniester as well as the Danube should have delivered such a vast surplus of freshwater that the Black Sea's lake swelled in size until it could pass the surplus on to the Mediterranean through cataracts in the Bosporus and Dardanelles.

Consequently, when the global sea level rose at the end of the last glacial cycle to the level of the top of the cataracts, there would have been only a wide body of freshwater for it to enter and not one shriveled by evaporation. To Ryan's knowledge no expert had ever doubted that meltwater, even in an arid regional climate, would have been sufficient to keep the Black Sea lake filled right to the lip of any outlet.

And so the curiosity of Ryan the daydreamer and Pitman the pragmatist, catalyzed by salt pans on the floor of a former sea, by the spectacle of the Gibraltar waterfall, by the presence of sudden faunal replacements, by the cutting of deep canyons, and by the rise and fall of the global sea level, subsided. Reading the Black Sea expedition reports had been sobering. Although the necessary conditions for a flood had emerged from the data, they were apparently not sufficient to have created one.

The charge of John Dewey to find Noah's Flood was mostly forgotten for two decades. Then one day the Iron Curtain came tumbling down.

CHAPTER ELEVEN

Aquanauts

O UT of the blue a fax arrived at Columbia University from a dispatcher in the former Soviet Union with a message that was absolutely unambiguous: The surface of the Black Sea lake had once been deeper than the level of the contemporaneous global ocean—and not so long ago at that.

> Bulgarian Academy of Sciences
> Institute of Oceanology
> Varna, Bulgaria
>
> March 19, 1993
>
> Dear Mr. Ryan and Mr. Pitman,
>   I have received Mr. Wegner's letter that informs about your interest in the Black Sea. I'd like to assure you, we have at our disposal convincing evidence that 9,750 years ago the Black Sea's level was at about 100 meters lower than today's level. This phenomenon had disastrous consequences for the natural environment. . . . At last, I'd be very glad to establish a personal contact with you and to express my good will for collaboration.
> Sincerely yours,
> Petko Dimitrov

Unbeknownst to Ryan or Pitman, their embryonic hypothesis of a Black Sea drawdown and flood had made its way to another dreamer, one who had been able to advance the idea with new facts. Over the course of several communications Dimitrov reported on a series of sediment cores from offshore Bulgaria that he had dated using carbon 14. His twenty-eight separate analyses painted a canvas far more complete than those of prior researchers.

*Petko Dimitrov emerges from the hatch of his Bulgarian submersible
following his discovery of a deeply drowned beach*

Everywhere seaward of the modern coast out to a depth of four hundred
feet Dimitrov had encountered what he called a "washing of the sediments"
that he believed had occurred as the former Black Sea lake shrunk to expose
its bottom to the actions of rain, wind, and surf.

Dimitrov had gathered his "convincing evidence" when diving far off
the coast of Bulgaria in a manned submersible. Using a scoop attached to the
sub's mechanical arm that he could position and operate visually from the
vantage point of a tiny porthole, he had collected deposits of shells en-
crusted with algae protruding from a sandy rock ledge diagnostic of an
ancient and presently deeply-submerged beach. The fax announcing his
findings was remarkable for both the nine-thousand-year-old date of the
shells and the three-hundred-feet depth of the beach. These values meant

without any doubt that there must once have been a barrier separating a partially-emptied Black Sea depression from an external ocean. Wally Broecker's saw-toothed curve of the last drowning of the Barbados coral reef placed the surface of the external salt waters some two hundred feet above the fresh waters of the Black Sea's former lake when waves were lapping against its now-fossilized shore.

Dimitrov's disclosure was timely. A team of Russian scientists at the P. P. Shirshov Institute of Oceanology in Moscow was just then applying to the International Atomic Energy Agency to learn whether the radioactive fallout from the catastrophic explosion of the Chernobyl nuclear power plant in Ukraine had reached the Black Sea and was being incorporated into its organic-rich black mud. If so, these dangerous substances might contaminate the food chain of its seabed creatures. The Russian scientists hoped to mount an expedition aboard one of their oceanographic ships the following June. Needing financial help, they were looking for participation by Western scientists.

Ryan and Pitman had been searching for such an invitation since the arrival of Dimitrov's letter. Anyone sampling the offshore of Ukraine and Russia with sediment cores to track radioactive cesium and cobalt could use these same sediments to confirm or refute Dimitrov's evidence of nine-thousand-year-old beaches now deeply submerged by seawater. Ideally one would sample not just the near-shore regions of the Black Sea where the radioactive substances might be most concentrated but other locations far out from the coast where, during the last glacial cycle of the Ice Age, rivers had cut their streambeds into an emerging arid landscape. Finding the now-buried streambeds would require geophysical tools towed behind the survey ship just above the seabed to penetrate the mud cover with sound waves.

Ryan had overseen the building of such an instrument back in 1980 and had towed it on miles of electrical cable to take the first sonar pictures of the wreck site of the *Titanic* on the bottom of the North Atlantic abyss. Ryan had lost the prototype profiler when trying to bring it back on board in a gale.

Hearing about the proposed Russian survey from a colleague of Ryan's and hoping for exposure to potential customers, the Datasonics Corporation in Catamut, Massachusetts, offered to lend the Lamont scientists a brand-new portable sonar instrument it had perfected that was well beyond the capability of Ryan's original. Word came back from Moscow overnight that not only would the Americans be welcomed aboard with their fancy profiling gear but that the scope of the survey would be expanded to sample the drowned shorelines and riverbeds using renowned Russian coring technology.

With invitations and clearances in hand, Ryan and Pitman hired a twenty-year-old summer intern, Candace Major, an undergraduate geology student at Wesleyan University in Middletown, Connecticut. They instructed her to track down monographs on Black Sea and Caspian Sea mollusks and gastropods in the Peabody Library at Harvard University. Thus, in the midst of preparing for final exams, Major undertook a self-taught crash course in fossil identification. It would be her job to give the others a knowledgeable interpretation of the ancient seabed environment through the identification of the seashells.

The three Americans packed their clothes and set off for an oceanographic laboratory set on the Black Sea's eastern edge at the foot of the Caucasus Mountains. At the end of their journey a dusty serpentine road that had been weaving in and out of endless mountainside stream valleys led to a portal in a fence mounted on a frame of pipes. A toothless old lady gatekeeper directed the visitors to a rustic cottage set in a grove of pine trees abutting an embayment on the verdant sea. Scattered about under the spruce canopy were a number of single-story buildings—residences for the staff, laboratories, and offices for the oceanographers. The only tall structure, reachable along a path to the shore, was a dacha erected in the previous century. Made of masonry limestone and covered with stucco and peeling chips of beige paint, its elegance was still visible through the obscuring film of decay. Inside an entrance hall vaulted up toward a decorated ceiling while encasing a central staircase that split and doubled back to the upper balconies. In the director's office on the second floor, Ruben Kos'yan was barechested in his trunks. His hair and trunks were still damp from an afternoon swim. Tall and with a tanned muscular frame, he tightly gripped each hand that he shook, a broad grin proclaiming a pleasure in the company of his guests from Columbia University.

Kos'yan apologized that the sonar device, shipped ahead by airfreight to Moscow and then to be driven to his lab by truck, had not yet arrived. He offered assurance that its whereabouts were being monitored. The driver had twice been threatened by bandits but had managed to evade the highwaymen on each occasion. He expected to make his delivery by late afternoon the next day. If not, he would postpone his arrival for another day since it was too risky to be on the road after dusk.

In contrast to the heightened anxiety and danger lurking outside the scientific compound, the serenity inside nurtured an ambiance of detachment. A remoteness that may have been a hindrance in better times now shielded the secluded community from chaos. The institute was famous for its pioneering research into the technology of hyperbaric diving. Breathing mixtures of helium, hydrogen, and oxygen, and tethered to a life-support chamber, aquanauts were able to swim more than one thousand feet below the sunlit surface to accomplish tasks requiring the full dexterity of bare hands. To descend still deeper, the center also operated a manned submersible to transport pilot and observer to the bottom of the deepest canyons in the sea, with eyes pressed to portholes for sightseeing and fingers on a joystick to manipulate a remote mechanical arm and claw to collect samples. Many of the ingenious devices to take the explorer into the abyss were built in the institute's shops staffed by machinists and their apprentices. The laboratory was at the disposal of the American visitors, who toured the facilities, including gardens, orchards, and greenhouses tended by the scientists to grow produce to take home when the market in the nearby city of Gelendzhik was bare.

The home away from home for eleven scientists and an equal number of officers and crew during the next fortnight would be the *Aquanaut,* a vessel 110 feet from stem to stern. Originally built to trawl for fish, the ship had been converted for oceanography by the Shirshov Institute many years before and with little structural change. A week ago, its crew had spruced up its hull and deckhouse with a bright coat of fresh white paint.

On the *Aquanaut*'s aftdeck, in temporary disarray with equipment to be mobilized, Kazimieras Shimkus, the expedition leader, readied the coring gear. Except for duty at sea, all of Shimkus's professional life had been spent at this institute. Two perpetually moist blue eyes set in a ruddy, weather-beaten complexion gave him a visage of melancholy. Though of medium build, his physique was authoritative. He flung the heavy, long steel core tubes into chocks like wooden javelins in a sporting contest. This researcher was a specialist on the seabeds of the Mediterranean and Black seas. He had read many of Ryan's and Pitman's publications, and welcomed the opportunity to accomplish serious work with the Americans.

The compact scientific lab occupied a crypt below the forward deck. A bench ran all along the starboard bulkhead, providing a platform to split, describe, photograph, and sample the cores that the Americans hoped to take. The lab would become even more confining with the installation of the control and display modules of the American sonar. In the process of a quick inspection of the available facilities, Ryan was informed that the pail of seawater on the tile floor served as the sink.

For the next two nights until they embarked for the high seas, the Americans slept in the comfortable guest cottage. Each evening they were joined by Kos'yan, Shimkus, and a few other scientists for a meal of poached fresh fish, potatoes, salad, and a delightful local wine. The banter at dinner in-

Walter Pitman (right) *and Kazimieras Shimkus* (left) *launch the CHIRP sonar from the Russian research vessel* Aquanaut

cluded toasts for friendship, health, and good fortune in the expedition ahead.

The truck from Moscow transporting the sonar did not pull in until the final hours before departure. As Kos'yan signed the manifest in front of a complaining driver, Shimkus pointed to a couple of shiny bullet holes in the passenger door. The crew transferred the heavy shipping containers from the truck to the foredeck of the *Aquanaut*. The part of the device that emitted the sound waves to probe the seabed looked like a giant pollywog with fins on its tail. The length and weight of an adult porpoise, this so-called fish would soon swim, attached to an umbilical cord, about ten feet away from the side of the ship and at an equivalent distance below the waves. The transmitted sound resembled a high-pitched birdlike warble. The Russians called the novel sonar the CHIRP.

Late that first night at sea, after about six hours of steaming westward from Gelendzhik, the *Aquanaut* approached the place where the ancient Don River might once have flowed on its way to the Ice Age Black Sea lake. Shimkus slowed the ship to the speed of a slow walk, a standard procedure when lifting a towed device from deck and submerging it in the sea. In less than a minute the fish was in its medium, streaming steadily alongside and beneath the *Aquanaut* without any observable yaw or pitch.

Down in the ship's laboratory, crammed with excited onlookers, Pitman activated the transmitter. Its song resonated through the thick steel wall of the ship's hull, repeated four times every second with an identically configured score. Faces of wonderment turned toward the video monitor of the CHIRP computer. It began to fill with columns of thin lines, laid side by side to build the two-dimensional profile of the internal structure of the seabed. It took but a few minutes to adjust the volume and time delays, and to optimize the graphical representation. Everyone recognized that they were looking for the first time at a dynamic color CAT scan of the seabed.

Immensely grateful that the equipment not only survived its perilous journey from Boston to the Black Sea but was even working, Pitman recommended that the ship's speed be increased to five knots for the actual survey. With the CHIRP now dunked, they would zigzag across the drowned path of the ancient Don River that had passed through the Kerch Strait to the shore of the once shrunken lake. At the instrument settings that had been chosen, the CHIRP displayed a thirty-foot-tall cross-section of the seabed. The uppermost three feet of the sediment cover was remarkably homogeneous. Its transparency to acoustic energy confirmed that it was water-saturated mud deposited far below the turbulence of surface waves.

The homogeneous mud rested on strata that were strikingly different, even to the neophyte. The underlying strata contained strongly reflective internal layering and many bright pockets that the Russian experts believed were the signature of trapped methane, a gas commonly produced in coastal marshes from the decay of terrestrial plants. The individual reflectors took on multicolored hues on the computer display and signaled an environment strongly affected by disturbances from wind, rain, and surf.

Barely an hour into the run, the ship crossed the ancient streambed now deeply buried beneath the homogeneous mud. Looking into the sub-bottom

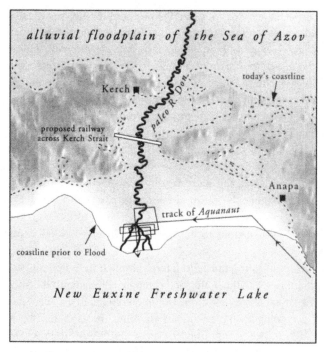

The Kerch Strait and the path of the ancient Don River

the scientists could see that before being drowned and smothered by the mud cover, its banks had stood fifteen feet above its streambed and its width exceeded a quarter of a mile. Without the CHIRP no one would have known the ancient river's whereabouts except through chance probing by expensive offshore drilling. Those aboard the *Aquanaut* had no doubt that the topmost homogeneous sediment had settled on the seafloor only after the channel had been drowned by a rising sea. Shimkus remarked that the image on the CHIRP monitor reminded him of the cross-section of the type of river that meanders in tightly curving loops across a floodplain before arriving at a coastal delta. He went on to say that he had never seen anything like this before on the bottom of a sea.

Ryan withdrew from the group around the CHIRP for a catnap in the top bunk of the cozy two-person cabin he would share with Pitman for the next ten days. He knew from prior expeditions that someone has to retire from the action so as to be ready to relieve other team members exhausted from the first shift of work. Alone and in darkness, he tried to clear the rush of excitement racing through his mind. So far the journey had exceeded his wildest expectations. Not only had the reception by Kos'yan at the Shirshov Institute been exceptionally friendly but from first impression Shimkus and his assistants seemed to be fully engaged, both emotionally and intellectually, in the search for drowned shorelines. The Russians brought to the joint enterprise a wealth of practical experience and a virtual treasure chest of local knowledge of the Black Sea geology. The discovery of a buried channel with all the characteristics of an ancient terrestrial river in the first two hours of surveying had to be more than fortuitous.

But if that was the case, why had the hypothetical flooding event eluded all previous researchers, both Russian and American? Could it simply be that the others had lacked the necessary instrumentation? On the other hand, might the newcomers be off on a false path based on illogical deduction, having overlooked something of vital significance to the traditionalist but somehow not yet obvious to the neophyte? Time would tell. Ryan drifted into sleep.

Pitman woke his roommate to inform him that the ship had passed over yet another buried channel, this one certainly belonging to the ancient Don River. The lateral levees had stood out in bold relief in the subsurface of the seabed, according to the sound waves. After the crossing, the ship had turned south to head away from the coast and toward the shelf edge. The bottom had deepened as expected. The CHIRP monitor showed that beneath the homogeneous uppermost sediment there was erosion surface that had beveled off the top of the deeper, highly reflective strata. This surface must have been land when the lake was shriveled.

Over the next forty minutes the CHIRP monitor received the full attention of Pitman, Ryan, and Shimkus as they traced the ancient land surface seaward to a depth greater than five hundred feet. There the bottom fell away at a rapid rate into the head of a submarine canyon. Along the way to the canyon the underlying reflective layering revealed many more bright pockets of trapped gas presumably being released from still-decaying plant material of former lagoons and swamps. One vent was caught in the act of discharging a plume of bubbles. Although minuscule, like those in a just-opened can of soda pop, the rising column of methane was visible on the computer screen.

Such plumes can be very dangerous because the charged gas, if sufficiently voluminous, drastically reduces the density of the seawater and hence the ability of a ship overhead to stay afloat. Drilling rigs and barges have been known to suddenly capsize, catch on fire, and sink with no survivors when a drill string inadvertently punctured a shallow pocket of marsh gas.

In just one twelve-hour set of zigzags across a suspected prehistoric river, the image of a shrunken Ice Age lake had materialized into tantalizing reality. To Shimkus, seeing was believing. The benches and terraces more than 450 feet beneath the *Aquanaut* could be nothing else but relict shorelines beveled by the pounding surf. The next goal was to get his hands on the actual beach sands.

Chapter Twelve

Immigrants

DESPITE his fascination with the CHIRP imagery, Kazimieras Shimkus felt an obligation to conclude the initial survey south of the Kerch Straits and steam west past the Crimean peninsula and on to the broad offshore shelf of Ukraine. It was time to begin his investigation of the radioactive nucleides released from the Chernobyl accident. Russian colleagues had already used sensors attached to moorings to detect an alarming concentration of dissolved cesium 137 in the water. This raised a concern of serious environmental danger since this soluble alkali metal would still retain a significant radioactivity well into the twenty-first century.

Shimkus's plan was to take the first core from the seabed in intermediate depths belonging to the mid-shelf and then head landward toward the mouths of the Dnieper and Dniester rivers where the contamination from Chernobyl had presumably arrived. Additional cores would be obtained every twenty miles or so until they reached the shallower inner shelf. Then the ship would reverse course and travel farther offshore until it eventually arrived in water approaching one mile in depth.

The CHIRP once more displayed the ubiquitous erosion surface lying just a few feet beneath the seabed, always covered by a uniform layer of what looked like homogeneous mud. Although there was a lot of speculation about what might have produced such massive erosion everywhere they surveyed, the answer would be found in the cylinder of sediment returned to the deck inside the twelve-foot-long steel core device.

Shimkus lined the inside of its tube with a sleeve of thin Mylar, which would allow him, once it was back on deck, to pull the sediment cylinder from inside the pipe and place it on a metal trough to be carried down to the ship's laboratory. On the lab bench he would snip open the jacket to expose the core for observation. Candace Major was to slice the sediment cylinder carefully in two lengthwise and roll the slender halves apart. In one half she would note the sediment layering, its color, and any other visual feature before she started to take samples at regular intervals to investigate the entrained fauna—the mussels, clams, and snails she had worked so hard to memorize. Shimkus and two assistants would examine the other half of the split core.

Shimkus had a wand that the Americans had never seen before. It emitted a resonating electromagnetic field. He slowly drew the wand down his half of the split core. As the wand moved, it energized water molecules trapped between the mineral grains in the sediment that he was probing. The amount of water present registered on a small display. Shimkus called out the numbers and the distance of each measurement referenced to the top of the core, and his assistant wrote them into her notebook. Shimkus reasoned that if the lake bed had emerged into open air in the past, its original moisture would have been evaporated away by the hot sun. The former sea bottom should have turned to terrestrial soil. Upon resubmersion, the soil should absorb only a small portion of new water, its pore spaces having been compressed by desiccation and by the trampling action of animals small and large.

Shimkus took samples that he weighed, dried in an oven, and then weighed again. The weight reduction was to be directly proportional to the amount of water lost in the oven—a value he used to calibrate his electronic instrument. Other samples were packed away to be returned to his laboratory to determine the amount of organic residues in the sediment, the composition of the minerals, and the concentration of radioactive fallout.

This first core created lots of excitement among the Americans. As the cylinder of sediment was slowly extracted from the core tube, Ryan immediately saw through the translucent sleeve the shells of species of mollusks he hoped would represent the earlier lake stage of the Black Sea. But his expectations were cut short. Only four feet of core emerged before the sleeve collapsed. Evidently the core tube had not been inserted to its entire length even though it had been lowered on its lifting cable at a relatively fast clip. Following the people carrying the cylinder of sediment to the workbench, Ryan looked at the CHIRP monitor and noted that when the coring device slipped into the bottom, the top of the buried erosion surface was four feet below the sea bottom. The massive coring tube, traveling at about five miles per hour, had stopped there as if it had hit pavement.

Sure enough, after Major sliced the cylinder in two and separated the two

*Candace Major discovers a Mediterranean immigrant brought
into the Black Sea by the flood*

halves, Ryan could see at the base of the glistening moist olive gray mud an
inch of pale-colored sand held together by hard, dry clay. Disappointed as
he was at having only scratched his objective, he knew that this dense
material, which had resisted the core barrel's penetration, represented an

entirely different environment from the bottom of a sea, despite the fact that it now lay in water 225 feet deep.

The mud was littered with the shells of mussels, many the size of a postage stamp and some as big as a house key. Their shell walls were very thin. In slicing the core, Major's knife met little resistance, only a faint crunching sound like breaking an exceedingly delicate glass Christmas tree ornament. The inner side of the shell was as iridescent as when the mollusk had been alive. It glistened in the overhead lights, revealing a luster of mother-of-pearl, royal purple, and sunset orange. *"Mytilus galloprovincialis,"* Shimkus called out just as Major was about to make her first identification. "No. . . . Well, I'm not so sure," she responded tentatively. "It could be *Mytilaster lineatus*. See, the valve is more oval than elongated."

Shimkus was taken aback. His authority was not often challenged and rarely by a woman—certainly not one so young and inexperienced. Yet with the specimen in his hand, turned around and around in suspenseful examination, he admitted that he had spoken in haste. These two species were practically cousins, and both were indicators of a saltwater environment. *Mytilus* is the common seashore mussel served up with lots of garlic, lemon, and basil as *moules à la marinière* in French bistros. Using a pair of tweezers, Major picked out a tiny clam, about the size of a dime in diameter, with both valves still attached at its hinge. Its external ribbing wrapped across the shell perpendicular to its growth rings. Pried open, the inside lip of pure white mother-of-pearl displayed teeth, straight and with subtle rounding at the tips, like those seen in a perfect film star smile. This was *Cardium edule*, another marine species common along the coast of the Atlantic and once a dominant species on the northern shores of Europe.

Cardium edule had appeared directly on top of the hard clay in association with a small rusty brown mussel called *Dreissena polymorpha* as well as another clam called *Monodacna caspia*. The *caspia* in the name of the latter species announced its significance. It indicated the presence of a rising salinity—the beginning of change from lake to sea. Although initially numerous directly on top of the dry clay, the brackish species rapidly became rarer as Major probed upward into the younger olive gray mud. Five inches above the hard clay they were gone and were replaced by more salt-tolerant marine species: *Mytilus galloprovincialis, Alba ovata, Parvicardium exiguum,* and *Retusa truncatula*. Finding the transition from fresh to salty beginning with the first sediments to be deposited across the formerly dry land was evidence that the Black Sea drowning was linked to an invasion of marine water.

As Major explored the core more thoroughly, working from its base or oldest sediments upward to its top and the most recent, she encountered twenty more marine species. Halfway up they were all accounted for. Ryan estimated that the introduction of new fauna had taken no more than a few thousand years. The remnants from the lake may have survived only a few centuries.

Ryan recalled that *Dreissena*, known popularly as the zebra mussel, had been the one to replace *Cardium edule* in the estuaries of Holland sometime in the 1930s when a single massive dike had closed the coastal embayment of Wadden Zee and had turned it into the freshwater IJsselmeer fed by the IJssel River. This lake is still being filled to gain new land for farming. In a reversal of the Black Sea flood, *Dreissena* had populated the floor of a sea turned into a lake in overwhelming proportions, replacing the existing fauna that were extinguished by the freshening. Taking advantage of waterways dug across eastern and western Europe in the name of commerce, *Dreissena* had journeyed to the IJsselmeer and Amsterdam from a Black Sea refuge in the Danube River attached to the hulls of barges.

Every Dutch citizen would understand exactly what would happen in the Black Sea were the Mediterranean to break through Ryan's hypothetical Bosporus dam. If the dikes protecting Amsterdam were ever to collapse again, as they did so catastrophically in the 1960s during a monstrous storm, the marine immigrant *Cardium edule* would be the first colonizer of the drowned polders.

As the *Aquanaut* traveled northwest toward the mouths of the Dniester and Dnieper rivers, five more cores were taken, each in progressively shallower water. Every time the core barrel reached into the bottom, its tip bounced off the hard sub-bottom erosion surface. Only a few tantalizing scraps of sand and stiff clay were recovered, not enough to test the theory of a partly evaporated lake. On the way back south, Shimkus asked the ship's captain to zig for about twenty miles and then zag for another twenty miles in a pattern that he hoped would criss-cross over the paths that he suspected these two great Russian rivers had taken to the shore of the Ice Age lake.

At the eighth coring location, this time in 325 feet of water, Shimkus instructed the chief mechanic to pay out the steel wire to which the heavy tube was attached at the fastest speed the winch's drum would turn. The gamble worked. Although the core pipe was only half-filled when it reached the afterdeck, the bottom two feet of the recovered sediment cylinder consisted of a substrate entirely different from the overlying homogeneous olive gray mud. When the core was split open on the lab bench, something special was present. Working from the top downward, the knife in Major's hand cut easily through the mud. But when she reached the substrate that had resisted the core barrel's insertion, her hand trembled with exertion as she continued the dissection. She struggled to saw it, so Shimkus took over. With his arm flexing its muscular biceps, he, too, had difficulty. The last foot of near solid clay eventually yielded to his strength and crumbled.

The passage from mud above to dry clay below had contained a thin layer of gravel that was particularly intriguing. When it was washed in the bucket that served as the sink, a residue of broken shells remained. The shells were more than shattered; nothing was left but the robust hinge and occasionally the thickest part of the valve where it joins the hinge. The fragments derived from the freshwater mollusk *Dreissena rostriformis*. They were pure white and porous, bleached by the sun and partly dissolved by rain. The clay below the gravel was also informative. When Shimkus measured its water content, he discovered that it was only a fraction of the amount in normal sea or lake bed mud. Shimkus also found a few thin lenses of sand within the clay that contained plant fragments and woody material as well as whole specimens of the *Dreissena* fauna, with both valves still attached and unbleached. Shimkus pointed to the thin dark scum attached to the outside of the shells. "That," he said, "is algae. It means these mollusks lived in a shallow lagoon. We must be near the shore of the vast Ice Age lake."

Shimkus poked around further and extracted several tiny snails no bigger than a grain of rice. In technical parlance the snails were gastropods. Their hollow spirals looked like the swirls of a unicorn's horn. According to the Russian experts, this particular assemblage of mollusks indicated that the sands were deposited in the environment of a prior coastal delta, possibly a distributary channel or bayou lake or even a marsh. None of these species are found today beyond the mouths of rivers such as the Don, Dnieper, and Dniester.

While Shimkus described the fauna in his notebook, Major scraped smooth the crumbled clay at the base of the core that had resisted her attempt to slice it. She hoped that its internal layering would reveal its origin. A few minutes later she called for Ryan's attention. She found cracks that had filled with sand. These cracks had been opened when the originally moist and fine-grained clayey soil had dried out in the sun, causing it to shrink. Such features are common in the ruts of rarely trafficked dirt roads a day or two after a thundershower. The shrinking surface breaks apart into polygons with three to five sides that may measure one to ten inches long on an edge. Dust kicked up by passing cars may settle in the crack and eventually fill it. In this case a sand storm had raged across the parched landscape.

Major's next visual discovery was dark circular spots about the diameter of the eraser on the end of a pencil, staining the dry clay that had cracked apart from shrinkage. Probing with her tweezers she pulled out a curly string of fibrous material from one of these spots. Ryan instantly recognized something he had seen in the drill cores from the floor of the Mediterranean when it had dried out to become a desert. "Why, these are the fossil roots of plants!"

The plants had apparently grown in place while the ground was still wet. Then the soil had desiccated. The plants supported by the soil and moisture had died. The roots had been buried under layers of clean, fine sand with partings that sedimentologists call laminations. The steep dip and cross-cutting nature of these laminations revealed the action of windblown dunes of the type Jirí Kukla had played on as a schoolboy in Czechoslovakia. Not only had the prehistoric dust storms ravaged eastern Europe, but they had deposited ribbons of desert sand along the rim of the shrinking Ice Age lake. Its depressed surface was no longer just an inference drawn from the CHIRP reflection profiles or a hypothesis deduced from incised river channels and sudden faunal replacements. It was a fact confirmed by the coating of algae on the shells of creatures living in shallow lagoons at the lake's edge and by the bleaching of these shells when the lake shrank, the marshes dried, and their formerly moist, muddy bottoms cracked from desiccation and became sun-baked clay.

CHAPTER THIRTEEN

Close Encounter

THE *Aquanaut's* scientific team then moved into a much deeper area of the Black Sea. Their next target for sampling lay beyond the outer edge of the continental shelf at a sounding of a little over four hundred feet. Once again Kazimieras Shimkus ordered that the weighted core barrel be lowered to the bottom at full speed. The hydraulic winch screamed in compliance. The slackening steel wire threatened to jump off its outboard sheave, a sign that the half-ton device was plummeting into the watery world below them at a rate limited only by the resistance of the water through which it fell. Shimkus knew from experience that at this pace the bottom opening of the tube would create a pressure wave in front of it. That jet would most certainly blow away the uppermost few inches of bottom mud and prevent him from recovering radioactive contaminants that might have settled at this spot since the catastrophic explosion and fire at Chernobyl. He was more than willing to make the sacrifice, however, in order to find how far seaward the ancient land surface extended.

Neither Shimkus nor any Russian or Bulgarian geologist had ever expected the shoreline of the giant freshwater lake to have been so far from the present-day coast. Yet the erosion surface in the image scrolling across the CHIRP monitor showed no sign of retiring. For the last twenty-four hours Shimkus and his shipmates had been mapping three large submarine canyons that had been carved into the steep continental slope. One of his coworkers read the depths from the echo sounder and plotted them in succession along the survey track on a chart. They showed the canyons to contain an inner channel that meandered just like a terrestrial river. In order to see such details, the *Aquanaut* had to be navigated with utmost accuracy.

To do this, its position was derived from telemetry transmitted by a constellation of satellites in the sky.

In 1983, Bill Ryan had been among the first civilian oceanographers to navigate ships by the orbiting beacons that now comprise the Global Positioning System, sponsored by the U.S. Department of Defense and now available to the public. He had needed three six-foot-tall side-by-side racks of electronic equipment to decode the signals. One decade later this task was being accomplished in a battery-powered unit no bigger than a cigar box. Although the Russians had their own navigation satellites, they preferred the simplicity and accuracy of the American system. They had sent a fax to Ryan the day before he left for Moscow, asking him if he would bring the necessary equipment.

When split and examined in the ship's lab, the 5.5-foot cylinder of recovered sediment was practically the same as in the previous core. The top was soft, moist olive gray mud. The bottom was stiff, dry clay with sand lenses deposited in what had been a coastal region. Sandwiched between was a foot-thick layer of the same gravel seen before, composed entirely of fragments of the most resistant part of the bleached shell of the *Dreissena* zebra mussel. Candace Major took extreme care in sorting through the gravel for intact fossil shells but found none, only detritus showing signs of abrasion and bleaching. If the gravel represented a coastal beach deposit, surely some unbroken shells would be present.

This anomaly puzzled her, and she asked for Ryan's advice. He suggested that she take a couple of handfuls of the underlying stiff clay and sand and put it in the pail. They carried the bucket to the foredeck where they found a fire hose and took turns spraying water into the bucket at about the same force that waves dissipate their energy on a beach in a moderate surf. Within half an hour they had created their own gravel layer. It had a composition and texture identical to the gravel sandwiched in the core. That gravel was apparently not a beach deposit but a seabed that had been exposed and desiccated, had been washed in occasional strong rainstorms, had probably been trampled on by animals, and had been bleached in the sun. It was as firm as the stiff, dry clay from the bottom of the sediment cylinder that held thin sand layers with intact *Dreissena*. Those mussels had formerly lived in the quiet setting of a shallow lagoon or bayou lake that, in the process of drying up, had shriveled into puddles. Every place that had been submerged and wet had made the passage to dry land through an interim stage of being a shore with water lapping against it.

The stream of the hose simulated waves breaking on that shore. It stirred the fine clay particles into a muddy slurry, which steadily overflowed the lip of the bucket and dribbled across the teak decking to scuppers that emptied into the sea. As the clay disintegrated, the intact mollusks were liberated. They settled to the bottom of the pail. Gradually the slurry lightened in color and cleared as more and more of the clay washed away. The heavier shells remained. Major found that by scooping them into the palm of her hand and directing the spray of water on them, she easily broke them apart. All that was left at the conclusion of the experiment was the robust hinge and an occasional thick piece of shell wall. She spread the residue out on a towel over a hatch to dry. By dinner time her manufactured gravel was as white as snow. Following her inquisitive intuition, she had demonstrated that the real *in situ* gravel had been formed by a falling sea.

The survey expanded to include a strip of territory some twenty-five miles west to east and about half that distance north to south. The CHIRP profiler illuminated a ribbon of dunes following the 250-foot bottom contour. They were linear ridges about one mile in length and eight to ten feet in height. Where sampled by coring, the dunes were draped by a uniformly thick layer of olive gray homogeneous mud overlying shelly debris abraded to sand-sized particles. Pitman realized that if the sea had gradually drowned the terrestrial landscape, the dunes would have washed away or at least degraded their shapes as they moved from land through the dynamic surf zone to become a seabed. Loose windblown sand would readily submit to the beveling process of beach migration. And even if the dunes partly survived the submergence, he argued, they would have been leveled by the motion of sea waves.

However, the sonar imagery spoke for itself. The dunes were pristine. The mud that smothered them was no more abundant in the troughs than on the crests. Only a very abrupt drowning could have accounted for their remarkable preservation.

The eastern sector of the survey region contained a broad terrace. At its outer edge, the erosion surface at last disappeared. As the *Aquanaut* crossed the seawardmost beach of the former lake and ventured into a deeper portion of the Black Sea, the seabed below had never emerged and been subject to washing of rain and surf. Its sediment carpet contained no gravel, no plant debris, and no dry soil—only soft, water-saturated mud and clay representing an environment that had always been wet.

To locate the lake's edge when it had shrunk to its smallest size, the *Aquanaut* crew turned back toward land and proceeded slowly until the erosion surface was once more visible on the CHIRP monitor. A short distance beyond where this reflecting interface was distinct and unquestionable, the scientists took one more core. Meeting their prediction, it contained the same shelly gravel and desiccation features as encountered where the erosion surface had been previously sampled. In this fashion the lowest shoreline was demonstrated to lie somewhere between 520 and 550 feet below today's sea surface.

The expedition had been a fabulous success, exceeding even Petko Dimitrov's estimate that "the Black Sea's level was at about three hundred thirty feet lower than today." Before accepting Dimitrov's age of the beaches at face value, however, the *Aquanaut* scientists insisted that the newly obtained core material be independently dated. Anticipating that they would conduct their own carbon 14 measurement in shore-based laboratories back in the States, Major and Ryan had carefully picked from each core the mollusk shells they found in the base of the olive green mud lying directly on the dry soil. They took four or five specimens from each core just above the wet/dry contact, giving preference to those with their two valves still attached. These creatures were the first settlers of the drowned landscape. The timing of their appearance at each coring site would document when that particular piece of land had sunk beneath the sea to become habitable for sea life. Since the transect of cores extended from beyond the outer edge of the shelf well in toward land, reaching the middle and inner shelf, Major and Ryan's samples would allow them to time the progressive immersion of the entire margin of the Black Sea as its shore marched landward. A gradual drowning as the continental ice caps melted, like that of the Barbados coral reefs, predicted a large span of dates between the cores at deep and shallow settings. In that scenario the level of the Black Sea had to have risen in tandem with the level of the global ocean once the glacial ice caps started melting. According to Wally Broecker's saw-tooth curve, the rise of the coral bathtub rings encircling the island of Barbados even at its fastest rate would never have exceeded six feet a century.

On the other hand, an abrupt flooding from a ruptured dam between the Black Sea and the Mediterranean could be tested by its prediction of an entirely different scenario. Such an unbolting of the Bosporus sluice gates would have led to an inundation of all the Black Sea's exposed land lying below the level of the global ocean at the time of the dam's collapse. The sea would have poured into the hole at a rate limited only by the size of the breach. Under the likely circumstance that the barrier would quickly wash away from the energy of the thundering cascade, the lake might rise by six or more feet every week, not every century. The carbon 14 ages of the shells of the first immigrants might resolve the question because in the circumstance of an abrupt flooding, the radioactive decay of the carbon atoms inside the individual shells would give the same date for the first immigrant fauna at all of the *Aquanaut*'s coring locations, deep and shallow.

The carbon 14 method had the precision to distinguish between rapid (a few years) and gradual (a few thousand years) scenarios once the hand-picked shells were delivered to a dating facility. Ryan had one already in mind. A former Columbia graduate student had built such a facility at the Woods Hole Oceanographic Institution with a state-of-the-art Accelerator Mass Spectrometer, thanks to a multimillion-dollar grant from the U.S. National Science Foundation.

WITH the CHIRP profiler performing so splendidly, Shimkus decided to extend the expedition a couple of days to see if the erosion surface could also be found along the margin of the Black Sea at the foot of the Caucasus Mountains. He reasoned that it was too late to apply for clearances from higher authorities, so he just proceeded, counting on the political climate that had lightened in the wake of *perestroika*.

The unique vantage of looking from the ship's rail at the distant pine-forested foothills rising from the sea gave Ryan and Pitman the opportunity to see firsthand the entrenchment of rivers caused by a substantial lowering of sea level. Every few miles along the coast a stream descended the mountainside in a deep gorge cut into the limestone and sandstone bedrock. But the gorge did not diminish its relief as it arrived at the shore. It just carried on with increasing proportions into the sea as if the sea had never been there. As a result, the water's edge was dented with valleys that made ideal refuges. The harbor of Colchis, Jason's destination in his mythological search for the Golden Fleece, was such a protected cove. Gelendzhik and Novorossisk were two others. Like the Sirens who beckoned Odysseus to drive his ship onto the rocky coast of Sicily, the Caucasus embayments tempted Shimkus. While exiting one winding anchorage, a famous cruise ship had sunk, and more than a thousand passengers and crew were lost. Its submerged wreck baited the expedition leader into a careless decision. Over several hours under the cover of darkness, the *Aquanaut* pirouetted and danced across that harbor exit until the CHIRP screen lit up with a ghostly image of a hulk, lying on its starboard side, with torn-away smoke stacks serving as grave markers. Ryan got the same feeling in the pit of his stomach as he had when towing his sonar "fish" over the *Titanic* ten years earlier, trespassing on sacred ground.

Apparently the KGB felt the same way. Agents had been tracking the

Aquanaut for more than a week, suspicious about radio messages they overheard describing the natural gas pockets and seeps discovered with the sonar. The zigzag tracks to find submerged riverbeds had been interpreted as spying for buried communication cables. Americans had never before been allowed to explore the Russians' territorial sea, especially with the fancy equipment the expedition leader had been raving about. And what was going on in the harbor?

Once back at the home port pier and after greeting his laboratory director, Shimkus appeared agitated. When Ryan inquired, he learned that the KGB had begun to interrogate the director about the unscheduled surveys. Shimkus was asked to turn over to the authorities the computer tapes acquired during the operations with the CHIRP sonar. Ryan volunteered the tapes of the suspect harbor survey but not the important ones from the program that had been approved by the Russian authorities. He was even willing to tear some pages from his notebooks, describing the last days' work, as long as they didn't want the seashells inventoried by Major in hundreds of small vials.

In Moscow overnight before returning to New York, the Americans checked into a four-star international hotel managed by Aeroflot for its foreign investors. Ryan and Pitman shared a room. They entered using a plastic card, coded at the front desk and inserted into the door lock. No sooner were they in the room than the lock clicked again and the door reopened. A huge man, the size of a professional football offensive lineman, dressed in a tailored business suit and carrying a briefcase, stared at the Americans and gazed around the chamber. Then in heavily accented English he asked politely, "Are you sure you're in the right room?" Without waiting for an answer he continued, "Excuse me. They must have made a mistake at the front desk." He stepped out, closed the door, and vanished.

Pitman, more astute than Ryan, sat on the bed, stunned. He pronounced, "He must be KGB! I am sure of it. Look, he spoke to us immediately in English. Did you see how his eyes moved from us to our luggage, the just-opened closet, the bathroom, the bedside tables? He was trained to memorize in a glance everything he saw before we could put it away."

They had been invited to the Moscow Circus that evening. Pitman, feigning exhaustion, excused himself, but Ryan and Major could not resist the prospect of being entertained by flying acrobats, dancing bears, and the overall gypsy ambiance. After his solo dinner in the hotel's dining room, Pitman returned to his room, but his plastic key would not work. He took the elevator down to the front desk, where an adjustment was made. He assumed that this detour had been a precaution to prevent his stumbling on an intruder searching their possessions. Yet nothing appeared disturbed; everything was in place. Standard procedure?

At the Shirshov Oceanology Institution the following day, Ryan gave a scheduled talk to researchers from the Russian Academy of Sciences and Moscow State University. He described the new results while projecting transparencies of the CHIRP sonar images onto the screen. The reaction of his audience was polite but subdued. A few people whom he knew only by name and reputation engaged in the discussion that followed, but it was from the perspective of their own mapping and coring, as if the results just obtained did not exist. Finally, an elderly woman, whom he took as a specialist on mollusks and who had probably been the author of the monograph that Major had memorized, asked him what he would be doing with his samples. Ryan's interest was piqued. He responded that he planned to identify them carefully and then date the shell material.

"But you have a problem," she announced. "The shells belong in museums. They won't let them out of the country. They are a national heritage." She then volunteered that she and a colleague could do the identifications in her lab. What could be better than to have the renowned expert, herself, check over Major's preliminary shipboard results? The negotiation took only a few moments. A final written report would arrive in the fall.

Major had sampled two dozen cores every few inches along their entire lengths. From these samples, Ryan chose a small allotment, which was wrapped and delivered by taxi to the address he had been given. Everything else he bundled into a valise for the flight home. In the car ride to the airport, the host from the Shirshov Institution admitted that he had been retained and interrogated at Lubamova Prison about the *Aquanaut* data. He felt that the Americans would be at grave risk to rush their scientific materials out of the country. He recommended that they leave the valise with him. His institute would check it over and determine that it was legitimate. Afterward it would be sent by courier to London and then mailed on to New York from there.

Pitman agreed immediately, but Ryan, who knew that the theory of a Black Sea flood would collapse without the actual material to date its onslaught, thought they should bring it home. He was sure that everyone was overreacting. Forty-five minutes later, however, when unloading the trunk of the car, Ryan's wits took hold. He said good-bye to the container, wondering if he would ever see it or its contents again.

On the conveyer belt of the Delta terminal of John F. Kennedy Airport, Pitman's and Major's bags arrived promptly. But Ryan's big blue duffel with his clothes and borrowed satellite navigation receiver was nowhere to be seen. Making an inquiry, they learned that this bag would arrive later on an Aeroflot flight, and he could pick it up from U.S. Customs. Sure enough it showed up in a few hours with an Aeroflot baggage tag.

Back at Lamont-Doherty, Ryan returned the navigation device to its owner. The person from whom it had been borrowed turned on the display to see if the apparatus was still working. The LCD window always presented the last fix obtained before it had been shut off, held in memory by a battery. The location was not the dock at Gelendzhik where the chief mechanic had thrown the bow line to tie up his ship at the end of the expedition. That position had been overwritten by the coordinates of the Sheremetievo II Terminal north of Moscow. Someone going through and inspecting Ryan's suitcase had found the GPS receiver and turned it on. Had the treasure of data from the floor of the Black Sea been in the bag and not left behind with the host from the Shirshov Institute, it surely would have been confiscated.

Beachcombers

G LENN Jones arrived early at his office at the Woods Hole Oceano-graphic Institution one January morning. The narrow, naturally weathered cedar walkway led him over glacial boulders and through a grove of bayberry and holly to the McLean Laboratory, tucked away in a protected hollow of the blustery hilltop Quissett campus of the famed marine lab. The 1993 Christmas holidays had been frantic. Today he had a visitor. Jones was particularly eager to show this guest around his facility, to see his reaction, to chat, and to reminisce. His caller would be giving a noontime lecture. Jones had circled it on his calendar with a fat felt-tip pen.

Before his visitor arrived, Jones had to tackle the backlog of reports on his desk. One chore was to go over the galley proofs of a manuscript he had submitted to the British journal *Deep-Sea Research*. Jones was rebutting the idea that ultra-thin and alternating dark and light layers of sediment accumulating on the floors of inland lakes and seas could be counted to tell accurate time. The champion of this approach had been Egon Degens, the co-leader of the *Atlantis II* expedition to the Black Sea in 1969, who had moved on from Woods Hole to become a distinguished member of the faculty of biogeochemistry and marine chemistry at the University of Hamburg. There were many other adherents as well.

In the interim, Jones had overseen the design and construction of a multi-million-dollar device that measured time by counting atoms instead of sediment layers. He was inspired to gamble on new cryogenic vacuum technologies that might allow the dating of more than four thousand samples in a year, well over twice the production rate of any other facility in the world. Jones had been right. Next month would mark the two-year anniversary of routine operations at the National Ocean Sciences Accelerator Mass Spectrometry Facility. Its precision had steadily improved and was now so good that he could obtain repeatable results to within forty years for material that was ten thousand years old. This feat was not accomplished in the old-fashioned way of listening to the ongoing slow beta decay of naturally radioactive carbon 14, a procedure that could take a week of continuous monitoring, but by measuring the ratio of the stable carbon 12 to the unstable carbon 14 remaining in the sample.

In Accelerator Mass Spectrometry (AMS) dating, the carbon atoms in the sample were turned into a plasma. Stripped of their electrons, the nuclei were propelled down a long tunnel by 2.5 million volts of electricity until they reached a velocity of ten thousand miles per second. Powerful magnets, tuned for atomic masses 12 and 14, bent the beam into tight but ever-so-slightly-different arcs and directed the split paths to separate ultra-sensitive detectors for counting their relative abundance of the two nuclei. In this fashion tiny fragments of seashell weighing less than one one-thousandth of an ounce could be analyzed in five minutes. However, Jones preferred to take a little longer and repeat the measurement eight more times to improve the accuracy of the results.

"AMS is the way to go," he boasted. His machine was currently giving a higher precision than the best conventional lab in the world. In many disciplines of the earth sciences, resolving controversy came down to chronology. What was the timing of this event? Did it come before or after that event? He would soon tell his guest, "AMS allows you to answer so many questions that couldn't be answered before."

And thus his high-tech dates on some of the very same Black Sea cores used by Degens and the other counters of ultra-thin sediment layers demolished their conclusions. Correlative climatic and chronological implications derived from the assumption that alternating dark and light layers represented strict seasonal changes were pure rubbish. Nevertheless, Jones was a gentleman. His page proofs did not attack the previous research outright. The dismissal was polite and gracious. He checked over each paragraph of the proofs word by word, each table number by number. He was not after typesetting errors alone because equally important was any omission or confusion that might not accurately credit the scholarship of others.

The first object ever to be probed by the radioactive decay of carbon 14 was a precious specimen of acacia wood excavated from a ceiling beam of the tomb of Zoser, deep in the interior of the Step Pyramid at Saqqara, Egypt.

This sample, measured in 1947, gave a reading six hundred years too young for the alleged date based on astronomical chronology and king-lists.

Keeping to tradition, Jones had inaugurated his facility by redating the same wood. But this time he required only a toothpick compared to practically the whole joist that was destroyed during measurement forty-five years earlier. His new age, calibrated by tree rings that chart the temporal variations of carbon 14 production in the atmosphere, was completely consistent with the archaeological estimate—right down to the century. One could do no better since it was impossible to know how long in advance of the pharaoh's death the tree had been harvested.

Jones had a similar high confidence in his Black Sea sediment dates. In the Late Bronze Age, the volcano of Santorini in the Aegean Sea had erupted in a cataclysm ten times more energetic than the explosion of Krakatoa in the Indian Ocean in 1883. The detonations of Krakatoa were heard in Australia at a distance exceeding a thousand miles. A huge tsunami drowned the entire populations of coastal villages.

Santorini jetted a plume of ash high into the stratosphere where it was carried by jet streams southeast all the way across the eastern Mediterranean to the Nile Delta and northeast to blanket the floor of the Black Sea with a dusting of angular glass shards. The discharge of soot and aerosols was so enormous that darkened skies inflicted several consecutive years of abnormal global cooling. The frigid temperatures and reduced solar illumination affected tree growth. Stunted fossil tree rings dating to the time of the explosion had recently been found in Ireland and in the Sierra Nevada Mountains of California. Splicing the tree ring record back through time placed the Santorini eruption at 1626 B.C. plus or minus a few years and within the accuracy of AMS measurements on seeds of grains found in a storage jar in the Minoan village of Akrotiri, entombed under the pumice on the volcano flank. The values obtained by Jones from the remains of marine plankton in the black mud bracketing the ash had coincided with the same date.

While conducting this test Jones made an unexpected discovery regarding immigration of fauna and flora into the Black Sea following its reconnection with the Mediterranean Sea. This was the news he wanted to tell Bill Ryan, the guest who would be arriving shortly. Jones was slightly embarrassed. Ryan had contacted him back in the fall when his seashells had arrived from Russia as promised. They knew each other well. Jones was captivated by the mysteries of the abyss while taking the first course Ryan had ever taught at Columbia University. He learned of the Santorini eruption during one of

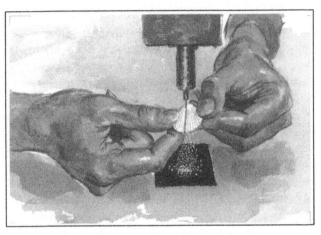

A few hundredths of a gram of powder drilled from the shell of Cardium edule *for a carbon 14 age measurement*

the lectures. He remembered tinkering with sophisticated instrumentation in Ryan's lab to try to measure precisely the size and shape of small particles transported far out into the oceans by wind and currents.

Jones had to tell Ryan that the shells had not yet been dated. They were still in the queue. He hoped that this afternoon the lab technicians could begin the process of scraping off the mother-of-pearl with the aid of a dentist's drill. The powder from tiny holes in the deeper substrate would be turned into graphite pellets before being loaded into the accelerator mass spectrometer. Perhaps Ryan could watch the preparation as a reward for his patience.

Ryan had been invited to Woods Hole to deliver the noon lecture, but he now had no punch line—the date of his alleged flood. The mollusk shells might refute the flood hypothesis and show that such a phenomenon had never happened. If so, it would be a former student who would disprove the hypothesis.

But Jones had some evidence of his own that was compatible with a rapid transition of the Black Sea freshwater lake into a saltwater sea. The sudden faunal turnover found by David Ross and Egon Degens back in 1969 had been dated with greater precision by the AMS method. The first arrival of the marine species had occurred simultaneously with the onset of the suffocation of the Black Sea by poisonous hydrogen sulfide at all depths—something never before suspected. This meant that the increase in salinity had been more rapid than previously thought and that the massive amounts of salt water pouring into the Black Sea had been deficient in oxygen. The page proofs about to be mailed back to England asked: "Does this revised timing for the initiation of anoxia [oxygen deficiency] compare favorably with the timing of the reconnection of the Black Sea with the Mediterranean?" Jones thought that it might.

Why had other scientists overlooked the linkage of saltwater influx and a rapid rise of salinity? They had been misled by the tiny aquatic algae named *Emiliania huxleyi,* discovered in the field of view of their powerful microscopes. This single-celled coccolithophore lives today in the skin of the world's ocean, where it is bathed in sunlight. The experts had found this microscopic plant in David Ross's cores in layers deposited long after the abrupt transition from the freshwater clay to the marine black mud. They presumed that the salinity had risen slowly and that *Emiliania huxleyi* had not populated the Black Sea until the salt content of the former lake had risen close to its present value, which is about one-half that of the open ocean salinity, a level that this coccolithophore could then tolerate. Jones had dated the first immigration of *Emiliania huxleyi* into the Black Sea as early in the seventh century B.C., almost five thousand years after the first entry of the marine water. He initially thought, as did the others, that the conversion from lake to sea might have transpired gradually. Only after tedious work in obtaining a large number of precise AMS dates to mark the moment of *Emiliania huxleyi's* immigration did Jones recognize the flaw in this logic. The appearance in the cores far above the first sign of Mediterranean spillage turned out to be coincident with the era when ancient Greeks established their first commercial foothold along the Black Sea shore. In the span of a single century, colonies had sprung up at Byzantium (Istanbul), Chalcedon, Apollonia, Istrus, Olbia, Tiritaca, Panticapaum, Phasis, Trapezus, Amisus, and Sinope.

Emiliania huxleyi is a passive surface dweller unable to swim. Thus it could never have migrated upstream from the Mediterranean and through the Dardanelles and Bosporus straits using its own power against the outflowing surface current once the passageway had been opened. And since it lived at the surface, neither could it have gone into the Black Sea via the deep countercurrent (the hidden river explored by the research vessel *Chain* in 1961). Jones therefore realized that this opportunistic plankton had only been able to make the trip as a stowaway in the bilgewater of Greek rowing ships, set low on the water, with open decks and no caulking. The historian Herodotus had described how schools of fish were kept alive on long voyages in the bilgewater and were then netted to feed the crew.

In other words, the mechanism of *Emiliania huxleyi's* entry into the

Black Sea of the ancient Greeks was as a hitchhiker, similar to the zebra mussel *Dreissena* that made its way up the Danube River and down the Rhine into the barricaded Dutch lakes in the late nineteenth century and more recently by modern freighter into the Great Lakes of central North America. The timing of the migration had nothing whatsoever to do with the rate of change of the water chemistry of the new homeland. The floating algae had been hanging around the door for eons but could get through only when mariners, driven by economic and political pressures, learned how to conquer the strong head currents.

Jones was good at debunking hypotheses, which was one of the reasons Ryan had chosen him to do the analyses on the seashells recovered by the *Aquanaut*. Jones would not be just a crank turner, inserting the carbon in one end of the accelerator and pulling numbers out of the other end. Both scientists needed each other to figure out what had really caused the marked environmental changes in the prehistoric Black Sea. If Ryan was on the wrong track with an abrupt flooding model, Jones would set him straight.

Ryan arrived at mid-morning. He was not as annoyed as Jones had expected when told that he would not have the data for the tale he would spin at noon. Ryan gave his talk without dates and without a conclusion. The forty-five minutes was spent presenting all the new observations. His audience was anxious to see the CHIRP profiles. They became engaged and argumentative. At the end of the talk Ryan made his prediction in the public arena. If the drowning of the former Black Sea lake had been caused by a sudden flood, the dates Jones would get in the next few weeks would all be the same. On the other hand, if the dates were distributed over several millennia—older from mollusks living in deeper water at the shelf edge and progressively younger from mollusks living in shallower regions of the seabed—then nothing catastrophic had taken place. The answer was in the mechanical hands of a robot that would begin the following Monday to drill away at the individual specimens of *Cardium edule*, *Mytilus galloprovincialis*, and *Monodacna caspia*.

In mid-February Jones pulled up a spreadsheet on the monitor of his office workstation. Bill Ryan's name on the document caught his attention. Jones scanned the headings of the different columns: Core, Water Depth, Depth in Core, Species Analyzed, Conventional Age, Reservoir Age, and Calibrated Age. The entries in the columns had all been made automatically by the computer that ran his AMS instrument. All the ages were the same, which pointed to a single event in the past, just as Ryan had predicted. But the age of the event was fascinating.

Jones saw immediately that he was looking at the same date he had obtained months earlier from his own Black Sea cores. Those samples had been recovered from the floor of the deep central basin and on the slopes, whereas Ryan's had come from the shelf. The age when the first immigrants arrived on the drowned shelf marked the onset of anoxia in the abyss. Jones had previously thought the consumption of the dissolved oxygen in the waters of the Black Sea had taken one or two thousand years once the Mediterranean first started to trickle through the Bosporus portal. A geochemist at his institution had calculated back in the 1970s that the preconditioning for eventual total stagnation might have required five or six thousand years. Yet Jones knew the importance of the new dates staring him in the face. The poisonous condition had arisen practically overnight. A conclusion in the proofs that he had returned to England for publication would come back to haunt him. The demolisher of other scientists' theories had now demolished one of his own.

Jones called Bill Ryan on the phone and began to chuckle. "Bill, have you tricked me?" he asked in an interrogating voice. "Have you sent me shells all from the same sample?"

"Why, no," answered Ryan. "Of course not! Why are you asking? What has happened?" Then Ryan inquired in a worried murmur, "Is there something wrong?"

Jones had confidence in the numbers. He read them into the phone. "The calendar ages are 7,500 years, 7,580 years, 7,510 years, 7,510 years, and 7,470 years [before the present; that is, before 1950]. They are from your shells sampled at 123 to 68 meters in depth [406 to 225 feet]. The one sigma errors are plus or minus thirty-five to fifty years. Statistically they are identical, just as you predicted in your talk several weeks ago."

Jones waited for the news to sink in. All he heard in reply was "That's great! But I wasn't expecting an age quite so young."

"Well, frankly, neither was I!" Jones replied. "They have been fully corrected and calibrated." Jones added, "Do you know what that age signifies?" Without waiting for Ryan's response, he volunteered, "That's the age of the onset of anoxia in all my cores."

"My God," Ryan exclaimed. "You mean that when all that salt water poured in during the big catastrophic flood, it just shut down the breathing of the Black Sea?" He paused. "I'm amazed, too. There is no way we mixed up in the samples, either intentionally or by accident. We were extremely careful not to screw this up!"

"Do you have e-mail?" Jones asked. "Well, here comes the file. Those are the numbers you are going to have to live with."

Ryan stared at the incoming data. Though exhilarated that the simultaneous AMS dates had fully met his prediction, they were not what he had anticipated. He had been expecting a drowning age much closer to Petko Dimitrov's date for the Black Sea beach. A number anywhere between nine and ten thousand years ago would have done just fine. Nine thousand years before the present was in the title of an announced talk he would give in May to the New York Academy of Sciences. Now it seemed that he had sprinted out of the starting block prematurely.

It was late in the day. Pitman was probably somewhere on the Palisades Parkway, driving back to his apartment in Manhattan from Lamont-Doherty. Ryan headed home for dinner to collect his thoughts and call up his friend later in the evening. This he did, but he reached only an answering machine. He left the message: "Walter, I have good news and not so good news."

When the phone rang later, Pitman said, "Tell me the good news first." He was delighted with the evidence of a sudden event. He had never doubted it. The uniform drape of homogeneous mud across the top of the sand dunes on the Ukraine shelf had convinced him.

"Okay, what's the bad news?" he asked, wondering how anything could be bad news. But at hearing the young age for the flood—in the mid–sixth millennium B.C. he was intrigued. After a pause to collect his thoughts, he said, "Bill, the bad news isn't bad at all. It shows that the shells on Dimitrov's beach had been out of the sea and exposed to the sun for two millennia before being drowned in the great deluge." He added, "The young age is going to make things much more interesting!"

During the previous academic semester Ryan had worked with Candace Major and Kazimieras Shimkus analyzing the CHIRP reflection profiles, drawing contour maps of the surveys, and selecting the seashells to send to Jones. Pitman had begun to inquire into the archaeological record. The new high-precision AMS date for when the Bosporus floodgate had broken open gave him a benchmark for dividing prehistory. Pitman knew that farming had been practiced in neighboring Anatolia and Greece for more than a thousand years before the date he was just given. A community dependent on the cultivation and seasonal planting of its fields and storage of the harvested grains would have made a substantial investment in its homeland villages. Expulsion by a permanent flood would have been close to annihilation.

"An exodus would have had the power to build myth," he explained. "Foragers might have simply trekked to another hunting ground, another stream, another forest, as if nothing major had happened. But farmers and herders would have had to take their seeds and animals with them for survival."

CHAPTER FIFTEEN

Back of the Envelope

THE dating of the shells belonging to the first marine mollusks to enter and colonize the drowned rim of the Black Sea gave Bill Ryan and Walter Pitman not only a precise age for the flood event but a means to begin an exploration of its overall magnitude. How big had the flood been? How long had it lasted? At what rate did the water rush in and the lake rise? How rapidly must one have fled to escape from drowning?

Seeking responses to such a range of new inquiry demanded knowledge of a number of boundary conditions. For example, in order to estimate the magnitude one needed to know the height of the waterfall, the width and length of the opening, the depth of the opening below the level of the inrushing water, and the shape of the basin into which the water was pouring.

The pressure driving the water from the Mediterranean into the Black Sea would have depended on the available hydraulic head. One useful way to think of the hydraulic head is to envision water pumped from wells or reservoirs up into a water tower. The height of the water in the tower determines the pressure at which it will be delivered throughout the township. For those homes on hills not far below the tower the hydraulic head is less than for houses down in a valley.

The velocity at which water flows depends on the pressure and any friction or turbulence that impedes the flow. Like a tap in a sink that is fully opened, the greater the pressure, the larger the volume of water that passes through per unit of time. Take, for example, the giant Hoover Dam built across the Colorado River in the 1930s. When finished and with its sluice gates closed, this barrier impounded about ten cubic miles of water in its Lake Mead reservoir. If one were to drill a series of imaginary equal-size holes into the seven-hundred-foot-high open face of this dam at different points below its lip, they would produce jets of varying velocity. The fastest outflow would be from the hole farthest down the face where the pressure of water behind the dam is greatest. The slowest would be from the hole just below the rim. That is why the large turbines that make hydroelectric energy are placed near the foundation of the dam.

Now imagine cutting a gash down through the dam from its rim to its base. Water rushing through this chasm is propelled by a summation of all the pressures from top to bottom. But as the flood continues, the surface of the reservoir soon falls, and the integrated pressure behind the outburst drops accordingly. The water in Lake Mead would escape in a few days and flow down through the Colorado River and into the Gulf of California, raising global sea level no more than a hundredth of an inch.

Now imagine that this gash was cut through the Bosporus dam. The discharge would not slow. There would be no drop in the head of water feeding the torrent. The water impounded in the Mediterranean, the exterior and interconnected Atlantic, Pacific, and Indian oceans, would be so vast it would be considered infinite in its availability. Any enlargement of the opening in the dam would only increase the rate of inflow to the Black Sea, and the flow would slow only when the level of the Black Sea rose to cover the inlet and decrease the effective height of the waterfall.

The height of the Bosporus dam was initially unknown. One might assume that its base lay at the level of the surface of the Black Sea prior to flooding. Though Pitman and Ryan had a good estimate of where that fossil shore line was today, based on the CHIRP profiling of beach terraces and the cores that brought back sands, gravel, and seashells, today's position was different from that in the past. Two things had changed in the interim. First, the seabed had become shallower due to the accumulation of sediment since the flood. The magnitude of this shoaling could be accounted for by the thickness of the homogeneous sediment seen in the reflection profiles. It averaged less than twenty feet across the entire region that was submerged. The second change was a deepening of the region flooded due to the weight of the new water. Specialists call this sinking "isostatic subsidence." Under the load of the new water cover, the earth's crust would actually sag, displacing some of the soft and viscoelastic mantle beneath it and pushing the latter out and away from the area where the load is applied. The sagging can approach a value equal to about one-third of the thickness of the water added. However, the earth's crust is semirigid and distributes the effect of the load unevenly.

Rough calculations performed by Pitman suggested that the preflood beaches were a little more than four hundred feet below today's sea level. To get at the level of the top of the dam Pitman needed to know exactly where the surface of the global ocean was 7,500 years ago, at the time the salt water entered the Black Sea. Luckily he did not have to go far to find a rather precise answer to that inquiry.

In 1988, Rick Fairbanks, another researcher at Lamont-Doherty, chartered a ship to explore the modern coral reef of the island of Barbados in the Caribbean. Whereas Wally Broecker had relied on corals uplifted in "bathtub rings" by the slow emergence of the island to date sea-level highstands between 50,000 and 130,000 years ago, Fairbanks concentrated on the most recent era, from the last Ice Age freeze, 25,000 years ago, to its present postglacial thaw. To get access to the modern reef Fairbanks had to work offshore in water depths too shallow for conventional oceanographic research ships. He therefore cast around for a boat with a reduced draft and a spacious deck for a drill rig. A candidate ship appeared—the *Ranger,* owned by the U.S. Naval Under Sea Command and based in New London, Connecticut. By coincidence the *Ranger* happened to be in the subtropical Atlantic conducting surveys for the Department of Defense and had a vacant spot near the end of its schedule.

Fairbanks neither owned the sampling equipment he needed nor knew how to use it, so he turned to two experts from the oil patch who had years of experience in the waist-deep-bayous of the Gulf Coast. They were able to mobilize a portable wireline drill rig from Shreveport, Louisiana, and weld it temporarily to the afterdeck of the *Ranger.* This strange marriage of southern roughnecks and Yankee sailors was epitomized by a Thanksgiving dinner at sea in which the holiday feast consisted of turkey and sweet potatoes prepared Cajun style.

Fairbanks's project had financial backing from the National Science Foundation but was functioning on a shoestring budget; consequently, he was able to afford only a few weeks of charter time. The installation of the heavy equipment and frustrating sea trials consumed the first ten days. Unlike the *Glomar Challenger,* which could drill in one spot for many days by positioning itself with computer-controlled thrusters, the *Ranger* had to be anchored on the side of the reef. This meant the tedious deployment of four monstrous moorings spread apart like the legs of a spider across several acres of steeply sloping bottom. Thick steel ropes stretched taut like strings of a giant guitar held the *Ranger* in place, unable even to twist or turn. The late autumn trade winds blew at twenty to twenty-five knots day and night. With each surge of the ten- to twelve-foot waves across the shoals—to smash onto the reef crest barely a hundred yards away—the *Ranger* would rise and fall, causing the steel lines to vibrate and threaten to tear the ship apart.

The effort to drill deep into the reef and trace its vertical growth during deglaciation remained fruitless until the Thanksgiving holiday when the first samples of coral were recovered. Eventually hundreds of feet of the reef were successfully penetrated and recovered in sixteen separate holes. The coral growth encompassed an interval in time reaching back from the Middle Ages to nearly twenty thousand years ago and the climax of the Ice Age. The coral consisted of a species of the common Elk Horn variety known as *Acropora palmata* that lives in water only a few feet deep. Fairbanks could be confident that at whatever depth in the boreholes this species was encountered, the surface of the ancient Caribbean had been nearby. All he had to do to track sea-level change was obtain as complete a vertical succession of the reef as possible and then return the specimens to his laboratory for dating. In the process he would be the first scientist to show exactly when and how fast the oceans had filled with the melting of continental ice sheets.

Broecker had ventured to speculate at the dinner in Paris with Jiří Kukla that the change was rapid, but until Fairbanks no one knew it could reach twelve feet in a century and that it would have slowed down and then speeded up in a series of fits and starts as the climate cooled and rewarmed at a pace swifter than anyone had thought possible.

In dating his coral samples Fairbanks used both the carbon 14 clock and a new method of Thermal Ionization Mass Spectrometry (TIMS). Fairbanks received help from a young postdoctoral fellow named Edouard Bard who came from France to carry out the mass spectrometry. Together they measured radioisotopes in almost one hundred sequential samples. The close spacing of the samples in the drill cores gave Fairbanks an age measurement every few centuries throughout the entire interval of deglaciation.

In addition to confirming Charles Maclaren's nineteenth-century guess that ice sheet growth would be "an agent capable of producing a change of three hundred and fifty feet on the level of the sea" (Fairbanks and Bard

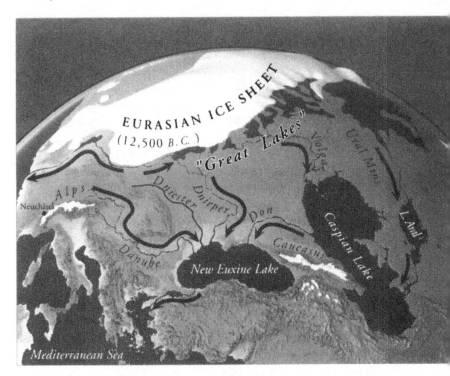

*The path of meltwater delivered from the Eurasian ice sheet
beginning around 12,500 B.C.*

measured four hundred plus or minus twenty feet), they showed that much
of the total ice cap melting had occurred in two brief and rapid spurts
separated by an interval lasting a little over a thousand years when the
climate had apparently returned to near Ice Age conditions.

The first of the rapid pulses was dated by Fairbanks and Bard as beginning 12,500 B.C. It was expressed as well in the *Aquanaut*'s cores from the
Black Sea by a deposit of light gray clay that Kazimieras Shimkus called
"New Euxine" (the Russian name for the Black Sea lake at its greatest expanse late in the last glacial cycle).

The first pulse of meltwater from the vast ice sheet covering northern
Europe and Asia was enormous. It fed dozens of "great lakes" that no
longer exist (for example, the Upper Dnieper, Upper Volga, Dvina-Pechora,
Tungusta, Pur, and Mansi). Their combined surface area plus that of the
swollen Aral and Caspian lakes dwarfed the single Black Sea Ice Age lake

by a factor of four to five. These lakes filled the sag in the Earth's crust caused by the weight of the huge ice dome. The lakes were dammed on their southern margins by a temporary bulge created when the weight of the ice sheet at its maximum extent pushed the soft interior of the earth aside. These "great lakes" swelled until one by one they breached the crest of the bulge and flowed southward to the Aral, Caspian, and Black seas. Each of the latter was strongly freshened by the voluminous discharge. One spilled into the next just as the Great Lakes of North America cascade from Superior to Huron, Erie, and Ontario. The widespread "New Euxine" deposits of the Aral, Caspian, and Black seas were the consequence of the impressive volume of the initial meltwater pulse.

The second meltwater spike beginning in 9,400 B.C., and seen vividly in the Barbados coral growth, never reached the Black Sea. This was the ice water that David Ross, Egon Degens, and most Soviet scientists thought had kept the Black Sea lake filled to its spill point. Yet somehow they had neglected a set of observations right under their noses—namely, when the ice caps over Britain, Scandinavia, Holland, northern Germany, Poland, and Russia had started to recede, the "great lakes" on their southern margins became trapped in the depressions made by the weight of the ice. The peripheral bulge directed the upper reaches of today's Dniester, Dnieper, Don, and Volga rivers to flow away from the Black Sea and westward across Poland and over Berlin to the North Sea.

The sideways escape route cut off the Black Sea lake from any further meltwater discharge. Jones's dates clearly showed that this event triggered the beginning of the final desiccation of the now isolated lake. As its surface evaporated down below its own outlet, the world ocean continued to rise. Using Fairbanks's dates from the Barbados reefs, Pitman could tell that already by 10,000 B.C., the Black Sea dropped below the level of the external ocean just as Petko Dimitrov had confirmed in his fax from Bulgaria. By 5600 B.C., its shoreline lay 350 feet below the top of the Bosporus dam. It was then, with the global ocean at 50 feet below today's sea level, according to Fairbanks's corals, that the trickle of salt water started, carrying the larvae that would become the Black Sea's first immigrants.

Pitman assumed that a barrier in the Bosporus dam had been breached, but he had little direct evidence. To gather some hard facts he traveled to Istanbul to consult with Celal Şengor, a friend of many years who had earned his Ph.D. in geology studying under John Dewey. Şengor introduced Pitman to two associates at the Istanbul Technical University. Mehmet Sakinç, a paleontologist specializing in bottom-dwelling plankton, had studied and

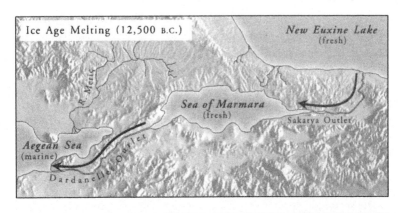

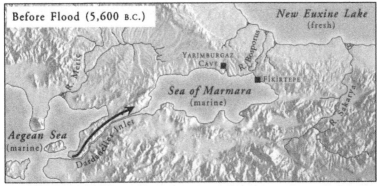

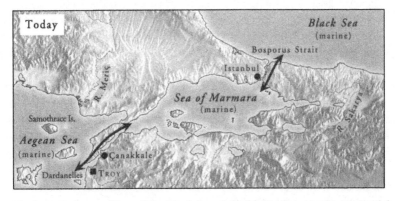

The historical connection of the Black Sea to the Mediterranean: the Ice Age lake drained through the Sakarya River outlet during meltwater discharge (top panel); the temporary isolation and drawdown of the New Euxine Lake prior to the opening of the Bosporus (middle panel); and the two-way exchange of water through both the Bosporus and Dardanelles straits today (bottom panel)

reported on the microfossils from new boreholes drilled in 1985 and 1986 from the floor of the Bosporus in planning for the Galata Bridge. The results from this drilling convinced Sakinç and coworkers that the sediments within the strait were no older than seventy-five hundred years and that conditions in the floor of the strait had been a mixture of fresh and salt water from the moment it opened.

The other associate was Naci Görür, who was familiar with sonar surveys conducted by the Turkish navy throughout the Strait of Istanbul (the modern Turkish name for Bosporus) and the Strait of Çanakkale (the Dardanelles). Both scientists knew of Pitman's hypothesis of a sudden Black Sea flood, and both thought it was supported by research the Turks had undertaken themselves.

Görür took Pitman to a restaurant in the small fishing village of Çubuklu along the Bosporus to meet in an informal setting Captain Hüseyin Yüce of the Turkish navy who was responsible for military oceanographic and hydrographic research. Yüce was well known internationally as an expert on the two-way current system in the Bosporus. Its floor had been thoroughly surveyed by echo sounding, and these surveys had revealed in detail the deep corridors encountered by the *Chain* back in 1961. Pitman learned to his surprise and delight that in addition to echo sounding, the Turkish navy had used seismic reflection profilers more powerful than the CHIRP to probe the configuration of the sediment fill in both the Bosporus and the Dardanelles. Furthermore, the navy had funded a student at the University of Istanbul to study these profiles for a master's thesis and to prepare maps displaying the thickness of the sediment fill. Although these new geophysical data were proprietary until the student's research was completed, Pitman was given permission to examine all the profiles with Görür and to make a tracing of three of the cross-sections.

What impressed both Görür and Pitman was that the bedrock surface tended to deepen as the survey progressed from the Aegean opening near Çanakkale northward through the Bosporus. The passage had been scoured down into contorted sandstone and shale some 350 million years in age. Several years later other profiles were made available and Pitman was able to compute the cross-section of the straits as if it were a pipe running from the Aegean to the Black Sea. Then, based on the depth of the gorge today and the level of the surface of the Mediterranean seventy-five hundred years ago, he plotted the varying hydraulic heads along the route. A simple formula from an engineering manual gave him a ballpark figure for the flow of water that would have passed through this conduit driven by the hydraulic head and inhibited by turbulence.

The flow was astonishing. The ten-cubic-mile reservoir of Lake Mead would have flushed through this orifice in less than a day, traveling at a speed of up to fifty miles per hour through the narrowest constrictions. The water rushing into the Black Sea lake would have raised its surface half a foot per day.

Pitman then thought back to the CHIRP profiles obtained aboard the *Aquanaut* as it surveyed the wide shelf west of the Crimea. There he had observed the seabed to be locally smooth and dipping seaward at the rate of no more than two feet per mile. Someone escaping the onslaught of the rising tide on this shelf would have had to travel, on the average, a quarter mile per day to keep up with the inundation. If fleeing up the very flat river valleys, they would have had to move half a mile to a mile per day. Though not difficult for someone on foot, in that nearly level landscape of a coastal setting it would have meant that one's village could have disappeared beneath the abyss in a matter of weeks. Of course there would have been advance warning as the opening of the dam enlarged from a trickle to a maelstrom. The seismic energy of the convulsing earth and the encroaching water would together have signified that it was time to pack one's bag and flee to high ground.

SOME months later Pitman was contacted by Bob Karlin, a geologist at the University of Nevada in Reno. Karlin had been to the Black Sea aboard the research vessel *Knorr* of the Woods Hole Oceanographic Institution in 1988, studying layers of sand and mud in cores taken all across its deep floor.

Pitman had lectured about the Black Sea flood at the University of California at Santa Barbara. Karlin got the news of a catastrophic event secondhand. Karlin was beside himself with excitement. Independently he, too, had come to the idea of a great "breakout" at Bosporus. He told Pitman that he had found a deposit on the floor of the Black Sea generated by an enormous underwater avalanche. Based on variations in its thickness and grain size from one coring location to the next, the material seemed to have come from a source directly north of the Bosporus's entry into the Black Sea where echo soundings on the continental slope had revealed a large submarine canyon leading to a radial apron of sediment. Karlin believed that the inrushing Mediterranean salt water triggered the avalanche and flushed the canyon. Its former fill was now spread out across a region several hundred miles in size, sandwiched between the light gray clay from the era of the Ice Age freshwater lake and the black jellylike mud of the marine sea with its new immigrant fauna and flora. This position in the *Knorr*'s cores put the avalanche right at the time of collapse of the Bosporus dam. Over the telephone Karlin asked Pitman, "But who would have believed in a flood based just on this single layer of debris?"

Who Was There, and Where Did They Go?

CHAPTER SIXTEEN

Anybody There?

WITH a location and date, and a calculation of the magnitude of the flood now in hand, Bill Ryan and Walter Pitman began to consider who might have witnessed it. Foremost was the strong possibility that the giant freshwater lake could have acted as a magnet drawing people to its rim for resources abundant there and perhaps sparse elsewhere. But who might these people have been? Indeed, who was around in Europe, Russia, Anatolia, or, for that matter, in any part of the Near East between the end of the Ice Age and the Great Flood?

The two geologists learned from their readings that the importance of a freshwater oasis in the middle of a desert as a catalyst for human cultural change had been proposed ninety years earlier by another American geologist, Raphael Pumpelly. For over seventy years Pumpelly traveled throughout China and Mongolia describing landforms and mapping rock formations. In 1904 near Ashkhabad on the fringe of the Kara Kum desert east of the Caspian Sea (in what later became Turkestan), Pumpelly uncovered signs of early farming at the edge of an oasis. Throughout his extensive travel he had noted that the climate in central Asia had become significantly drier in the wake of the last Ice Age. He wondered whether during this desiccation Stone Age hunters and gatherers had found themselves clustered together around the edges of the remaining water holes along with wild animals and plants. Perhaps in order to "conquer new means of support" these people took the crucial cultural leap leading to the genetic manipulations of plants and animals called domestication. Pumpelly formulated what has since become known as the "oasis theory of agricultural origins." Although it never found strong adherents in the United States, in succeeding decades it caught the attention of one particular Old World archaeologist.

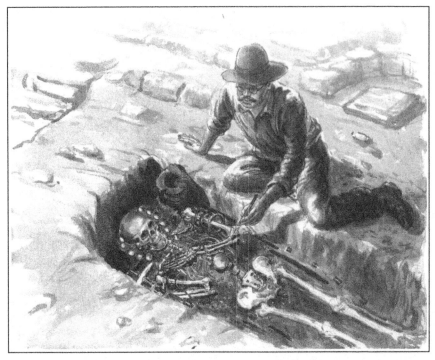

Vere Gordon Childe inspects a freshly opened Vinča grave
dating to the sixth millennium B.C.

This was Vere Gordon Childe, who would one day be proclaimed "the greatest prehistorian in Britain and probably the world." Short and with a slight build, he was shy, awkward, and homely. For six weeks in the summer of 1926, Childe and a boardinghouse friend went on an adventurous tour of archaeological sites in the Danube River valley. A shaven-headed White Russian general chauffeured them over rutted wagon paths in a large American roadster. The driver's instructions were not to lose sight of the Danube as it snaked across Hungary and Yugoslavia, descended rapidly via the gorge of the infamous Iron Gate gap through the Transylvanian Alps, slowed across the nearly flat plain of an ancient but now dried-up lake, took one last hairpin turn around the Dobruja massif in Romania, and emptied into the Black Sea.

On the way, the two vagabonds stopped in Belgrade to inquire about bones rumored to be abundant just downstream of the Danube's confluence with the Sava River. Because the Sava tapped its large watershed across the eastern flank of the Dinaric Alps, reaching practically to Italy and Austria, Childe thought this juncture might have served as a strategic trading post for the first settlers in Europe. In the course of his inquiry Childe came across a Danube culture in another red clay quarry, this one cut into a twenty-foot-high terrace of windblown soils ten miles downstream from Belgrade. The new culture had been nicknamed Vinča by its discoverer Miloje Vasic after the peasant village where the wagon path came to an end at the river's edge. Childe clambered onto the ledge overlooking the quarry pit and dug away with a penknife into the various layers of prior human occupation to determine the temporal sequence of the preserved artifacts.

The objects Childe inspected, sketched, and described during his six-week excursion led him to conclude that traders, prospectors, and even primitive metallurgists had come to the Danube and its far-reaching tributaries for the purpose of setting up villages and towns to support the mining of gold, copper, and tin for export to the Near East. He was particularly impressed by pins, earrings, and daggers with artistic styles that reminded him of Asian and Palestinian handiwork. He noted the presence of *Spondylus* shells from the Mediterranean seashore among the artifacts turning up in Vinča graves. Perhaps this was evidence of a long-distance trade. The Danube and its tributaries possessed a wide spectrum of natural resources to attract homesteaders whose farms were never far from abundant water. In his book *The Danube in Prehistory* he speculated that the early European farmers had purposely chosen to settle their villages and till their fields on the banks of rivers or near the shores of lakes.

In light of what we know today, Childe's view of the origins of farming were prescient. He labeled the development of agriculture "a revolution whereby man ceased to be purely a parasite and . . . became a creator emancipated from the whims of his environment." He followed Pumpelly's lead in proposing that the revolutionary active partnership with nature was triggered when, in response to a desiccating landscape, disparate nomadic groups gathered into a few oases where game, plants, and water were plentiful. Although Childe never placed this mythical Garden of Eden on a map, he was the first to demonstrate that Europe's farmers and their domesticates had come from Asia and thus had taken a route that would have brought them along the shore of the Black Sea. He pointed out that the oldest domesticated sheep in Europe is the descendant of an Asiatic species *(Ovis vignei)* native to Turkestan and Afghanistan.

Working in the era between the two world wars, Childe could only assemble a relative chronology for the Vinča folk. Believing the Vinča to be as young as the Bronze Age, he regarded his field of archaeology a tool of prehistory. He argued that the early record of Europe should be decipherable through the artifacts that he and others found in excavations. He extended his thesis even further and proposed that the languages spoken today had developed from ancestral tongues of the Stone Age and still embodied many concepts (such as gender for inanimate objects) from that earlier period.

A few years after his appointment as director of the Institute of Archaeology at the University of London in 1946, it became possible to date the wood, bone, shell, and other formerly living material dug up by archaeologists. When this method was applied to the prehistoric cultures of eastern Europe, much of Childe's relative chronology collapsed. As a result, his reconstruction of Europe's past was seriously challenged—and with it many of the concepts such as the oasis theory upon which his interpretations had been based. Childe became despondent at seeing his life's work apparently voided. He returned to Australia, his country of birth, in 1956 and within a year was found dead near Katoomba at the base of the thousand-foot-high Govett's Leap.

The radiometric clock did indeed show the Vinča to be thousands of years older than the Bronze Age that Childe had envisioned. Modern tree-ring-calibrated dates now place the Vinča as contemporary with the Black Sea flood. Less advanced farmers were in Europe even earlier, living in Greece, the Balkans, and out on the Hungarian plain. However, from the bones, tools, and pottery fragments found in relatively thin soil horizons, the less advanced farmers seem to have been mobile, relying to a great extent on exploiting wild resources. It is thought that they may have been semisedentary in contrast to the Vinča who built villages with houses of remarkable post-and-beam architecture on the same site over many generations. Yet it is notable that any camps of early farmers that have been found were situated along the banks of rivers with year-round availability of water.

In the winter of 1995, Ryan and Pitman traveled to England to inquire about the first farmers in Europe. At Cambridge University they visited Ian Hodder who had written a book entitled *The Domestication of Europe—Structure and Contingency in Neolithic Societies*. Hodder arranged for Ryan to report the new results from the Black Sea including the dating of its abrupt flooding in a presentation to faculty and students at the recently-opened McDonald Institute for Archaeological Research at Cambridge. Hodder was surprised to hear of a drawdown of the Black Sea lake by evaporation in the millennia before the Mediterranean salt water broke through the Bosporus portal. Like many of his contemporaries, he assumed that the climate ever since the end of the Ice Age had been more or less moist and warm. Paleo-botanists identified the sixth millennium B.C. as belonging to the "Atlantic" stage of postglacial climate with its wide distribution of deciduous forests rich in oak, elm, beech, and hazel trees.

Hodder took Ryan and Pitman into the library of the Department of Archaeology and showed them many books and monographs that he thought would be helpful for their investigation of who was in Europe at the time of the proposed flood. He then rang up Andrew Sherratt at Oxford University, one of the world's leading experts of Stone Age Europe and its landscape. Hodder arranged for the two Columbia geologists to meet with Sherratt in his office at the Ashmolean Museum.

Sherratt's greeting was especially warm and enthusiastic while he commented that the appearance of farming in the Levant and the Fertile Crescent in the tenth millennium B.C. predated that of Europe by at least two thousand years.

Once in Sherratt's cloistered office on a high floor of the museum, Ryan laid out on a light table the view graphs he had presented at Cambridge. One showed the Black Sea's vast Ice Age lake and its broad environs at the time of the maximum growth of the large continental ice sheets about twenty thousand years ago. Another showed the lake in a more shrunken condition with the deglaciation well under way. Sherratt, too, had made such reconstructions. He went to a stack of long, thin file drawers in which he kept his maps. He pulled half a dozen sheets and laid them out on the table over the view graphs. Although he cautioned that they were provisional, each had been sketched very carefully and filled in with colored pencil, showing a pattern of different climate zones and belts across Europe, southern Russia, the Black and Caspian seas, Turkey, and the Near East. The individual maps within the portfolio represented different slices of time, going back into the past at steps of two to three thousand years. At each interval not only was the vegetation portrayed, but also symbols marked each known site of human occupation at that particular moment in the climate development. Sherratt used the spores and pollen found in peat bogs and lake bottoms to estimate the relative abundance and types of the trees and grasses growing during each time period. Ryan and Pitman were familiar with some of the data but had never seen such a thorough compilation.

What stood out on the map from the sixth millennium B.C. was a vast forest and woodland extending throughout eastern Europe right up to the edge of the modern Black Sea. Sherratt described the gradual expansion of this forest from small Ice Age refuges in the Balkans and the Caucasus, beginning three thousand years before the alleged flood. The reconstruction was clearly incompatible with the widespread aridity that Ryan and Pitman believed had evaporated the Black Sea's giant lake.

When asked who was there on the European margin of the Black Sea, Sherratt responded that, in his opinion, the population would have been predominantly indigenous hunters and gatherers living in small bands and distributed in sparse numbers throughout the coastal woodlands and the interior forest. He doubted that they would have taken much notice of a flood. In continuing his description of the setting, he noted that these Mesolithic (middle Stone Age) folk were best known from the Iron Gates region of the Danube where the river breaks through a gap between the Carpathian Mountains and the Balkans. A people known as the Lepenski Vir folk had built a strange form of stone houses high on a ledge of a cliff on the Serbian side of the gorge through which the river passed in a series of rapids and waterfalls. Their buildings were trapezoidal and laid out in a semicircle facing toward the river.

Radiocarbon dating revealed the site had been used for hundreds of years before it was abandoned—around the time that the Mediterranean poured its salt water into the Black Sea. Sherratt went on to relate to his guests how Ian Hodder and Dragoslav Srejovic, the discoverer and excavator of Lepenski Vir, believed it had been used as a shrine for spiritual purposes. The only sign of domestication was the dog. A main source of nutrition came from catching carp, catfish, and sturgeon in the whirlpools below the ledges. Hodder saw at the Iron Gates what he believed was an expression of the human aspiration to control death and tame the wild in nature.

The kinds of fish identified by the presence of their bones scattered around the stone hearths of Lepenski Vir were noteworthy because they were not represented by any species that returns from the saltwater ocean to rivers for spawning. To Sherratt that was compatible with a Black Sea cut off from the Mediterranean by a barrier. Noting the lack of any Stone Age site on Sherratt's marked map in the vicinity of the present Black Sea coast, Pitman pointed out that today's coast was far from the ancient lakeshore. Any villages close to the shore of the freshwater oasis would now be underwater and far offshore.

The climate reconstructed by Sherratt was apparently warm and moist. Accordingly, woodlands had filled out the former Ice Age open terrain at the expense of grasslands. As far as humans were concerned, the Flood may have been a nonevent. From Sherratt's maps it certainly didn't seem that Pumpelly's and Childe's oasis theory had any relevance to eastern Europe.

-W-

In the late winter of 1996, Ryan and Pitman were in London to talk with the BBC about a potential film documentary on the Black Sea Flood. The show's producer introduced them to David Harris, the current director of the Institute of Archaeology of the University of London. Harris had just completed the editing of a book titled *The Origins and Spread of Agriculture and Pastoralism in Eurasia*.

Harris credited the Russian botanist Nicolae Vavilov (1926) and Vere Gordon Childe (1928) with the notion of "centers of origin" for the "agricultural revolution"—the where, when, and how-fast issues of first farming. Vavilov had proposed a handful of independent centers of origin for plant cultivation, which he considered to be "the homes of primeval agriculture." As recently as the 1960s and 1970s it was commonplace to adopt Vavilov's and Childe's proposition and consider multiple independent innovations, especially for the spread of farming from Asia and the Near East into Europe.

However, based on considerable substantive evidence, much of it collected only in the last two decades, Harris did not believe this was the case. For him the data showed that "agriculture originated independently only very rarely—possibly only twice—in the history of Eurasia" (first in the Near East and later in China). The earliest event had taken place in the late tenth millennium B.C. in what Harris called "rift-valley oases." He placed its location on the western side of the Fertile Crescent at a place such as Jericho, not far from the Dead Sea. In a thousand years the so-called founder crops spread northward to Anatolia and eastward to Iraq and Iran. Domesticated sheep and goats (the caprines) came a few hundred years later out of the Taurus and Zagros mountains of Anatolia and Persia.

Like his predecessor Childe, Harris believed that environmental change —particularly climate—was instrumental in bringing about the transition from foraging to agriculture. Excavations showed quite clearly that people lived in rather sizable villages for a substantial part of the year far in advance of the observed domestication of cereals (barley, einkorn, and emmer wheat) or pulses (pea, chickpea, lentil). The sedentism commenced about the time of the large meltwater pulse, seen by Richard Fairbanks, in the coral reef of Barbados beginning around 12,500 B.C. It thus occurred during the initial deglaciation when the entire Near East was warmed by shifting patterns of atmospheric and ocean circulation—far in advance of the Black Sea flood.

This period of post-glacial warming was interrupted by a precipitous and brief shift back to near-glacial conditions. Called the Younger Dryas in Europe, it lasted from 10,500 to 9400 B.C. The continental ice sheets ceased melting and in some localities (such as Scotland and Norway) began once more to expand, sending long tongues of glacial ice down mountain valleys and back into coastal fjords. The word Dryas comes from the name of an Arctic plant that is a member of the rose family and which appeared across northern Europe at the onset of the abrupt refrigeration. In the Barbados coral sequence, the Younger Dryas's cold and dry climate produced a drastic slowing of the rise of global sea level.

One of the first archaeologists who recognized a human response to the Younger Dryas was Andrew Moore, a fellow of Queen's College at Oxford. While away in Palestine, in the spring of 1971, he received a message from the director-general of antiquities and museums in Syria who was desperate for help in planning an emergency dig on the banks of the Euphrates in a region that would soon be flooded by a man-made reservoir. Moore responded with an immediate affirmative, unaware that this serendipitous phone call would make him the first to see in vivid detail the passage from foraging to farming. He hurried to acquire the necessary visa and soon arrived for a reconnoitering of potential sites to dig. Two unexplored mounds beckoned him. Choosing the one called Abu Hureyra he climbed forty feet to its crest. From whatever era it heralded, Moore could immediately appreciate its strategic placement on a bend of the Euphrates. He stared upstream at the notch on the northern horizon cut into the forty-million-year-old yellow flint-rich chalk by the river's Ice Age power. Looking downstream Moore thought of the inhabitants of the great city-states of Mesopotamia who had rowed in their boats up the Euphrates in prehistory to cut timber, hew rock, and prospect for metals in the mountains on his western horizon.

Working under the pressure of time, Moore and his contingent of archaeobotanists, archaeozoologists and paleopathologists had only two field seasons before the sluice gates of the Tabqua Dam closed and their trenches slipped beneath the water. As they sank their shafts from the top of the mound to its base, they passed through the remains of two successive villages stacked one upon the other. The lower and more ancient village was primitive. It had begun as a seasonal camp with a few outdoor fireplaces and then huts of reed, mostly circular or oval in shape and positioned above pits cut into the subsoil. These homes had been inhabited by hunter-gatherers who lived at this spot for a major part of the year except when they journeyed to open prairie land to hunt for gazelle. Deploying netting, brush fences, and the natural contours of dry streambeds they drove herds of these swift antelope into narrow enclosures (called desert kites) in which they selectively slaughtered only the young males. Moore could deduce the selective culling of the herds by the uniform small size of the bones in the carcasses brought back to the village and the immaturity of the teeth in Abu Hureyra's fossil garbage heaps. He was impressed by how sensitive the Stone Age hunters were in keeping the breeding stock intact, thus assuring large herds in future years.

Using salt and sun-drying these hunters, called Natufian by archaeologists, preserved the meat of their hunt for the months ahead. They also gathered a large variety of edible plants—more than a hundred individual species—to balance their diet. Moore discovered not only evidence of grain storage in bins hollowed from the limestone bedrock but also seeds and husks of plants carbonized by heat from ancient fireplaces and thereby saved from decay. He and his coworker Gordon Hillman sorted through the charred vegetable remains layer by layer using an ingenious homemade device to float the fossil plant debris in a froth of water. With fine-meshed sieves they also reclaimed thousands of fish bones and mollusk shells, which told them of the food caught from the river. One particularly valuable residue from the bucketloads of soil lifted from their trenches was charcoal. They sent the larger pieces off to Oxford for carbon 14 dating.

The ages of the charcoal spanned from 11,000 to 9,500 B.C., placing the first settlement of Abu Hureyra in the era of postglacial warming through to near the end of the Younger Dryas. The food residue revealed that the Natufians used sickles of carved deer antler studded with flakes of flint to harvest the natural stands of native wheat and rye. They reaped wild barley, lentil, and vetch, and the fruit of the hackberry, plum, pear, and fig tree, as well as the caper bush. Their diet of plants, fruit, and nuts, though coarse and stressful to their teeth and requiring back-bending labor with grinding stones, mortars, and pestles for preparation, was more than adequate for subsistence.

In the deteriorating climate of the Younger Dryas, however, the Natufian diet changed drastically and abruptly. The fruit trees of the neighboring forest or forest fringe retreated to refuges beyond the gathering range of the villagers. The dying of the woodland became apparent in the evidence of a sharply increased diet of cereals, grains, and grasses that took brief advantage of the disappearing shade. In turn the cereals became sparse and were replaced in the diet by more hardy plants such as clover that require substantial preparation to detoxify them before their pulp or flour can be eaten. Moore and Hillman called these the "fallback foods," which would have been suitable as staples only when the other major plant groups were no longer obtainable. Eventually all the valley-bottom plants were gone. To the excavators this indicated that the Euphrates no longer overflowed its banks even during spring floods. The diet evolved further to one almost completely reliant on fishing.

Moore found that in the youngest strata of the first settlement, representing no more than a few human generations, the fish bones and mollusk shells disappeared from the edge of the hearths. The Natufians either starved to death as a group or abandoned their village to forage elsewhere. Moore and Hillman finished their history of the first Abu Hureyra settlement sieving through a sterile soil that blanketed the deeper of the two superimposed villages: no more huts, no more tools, no more animals or bones, no more charcoal, no food residue at all as evidenced by the disappearance of village scavengers such as the mouse and sparrow.

The collapse of the Natufian sedentary way of life at the depth of the Younger Dryas desiccation and cooling was not unique to the middle Euphrates Valley but is found at other Natufian villages and camps across the Fertile Crescent (Jericho, Tell Aswad, Aïn Mallah, Beidha, and Mureybet) and even far to the south in the Negev highlands. With the elevated aridity a new hunting tool appeared—the delicately knapped arrowhead called the "Harif point," indicating what archaeologists believe to be an almost total reliance on desert mammals such as the gazelle, ibex, and hare. The artifacts and bone residues conveyed the message of struggling desert nomads who despite a large territory did not make it through the brutal Younger Dryas period. The primary cause of the more or less simultaneous desertion of the Near East by humans is today blamed more on aridification than cooling, although the latter was important and could be measured by tracking campsites as they were moved progressively to lower altitudes. The Dead Sea evaporated its saline lake by hundreds of feet.

When Moore began to process the food residues from the second and younger of the two villages superimposed one on the other at Abu Hureyra, he made another stunning discovery. Those who saw the same strategic advantage of that bend in the Euphrates River and built on the buried ghost town were already farmers. The newcomers distinguished themselves from their Natufian predecessors not only by the genetic alteration of the wild plants that Moore and Hillman could readily recognize but also new tools to till the soil, new materials to build their houses, new ways of cooking, and new practices for burying their dead. No one could tell for sure where the Neolithic farmers came from.

During their 1996 meeting in London, David Harris had explained to Ryan and Pitman the theory of the domestication of wheat. In genes and shape, cultivated wheat is quite similar to wild wheat. Domestication is the human creation of a new form of plant or animal—one that is identifiably different from its wild ancestors and extant wild relatives. The process requires changes in a few genetic loci producing heavier seeds and denser bundles of seeds.

Harris explained that under the pressure of aridity, ancestral wheat evolved through natural selection and produced small seeds (characteristic of most annuals) protected in a shell (glume) that was tough and could resist the loss of moisture. The seeds grew in rows in spikelets that were attached to the stem of the grass by a brittle hinge called the "rachis." The plant had developed a strategy to drop these spikelets from the stalk early and easily —perhaps by the rustling of small animals—so that the seeds within could get trampled into the ground and gain a foothold to start germination in advance of the harsh dry summers typical of the Near East. The Younger Dryas aridity accentuated this adaptive process. Humans had merely to harvest the ancestral wheat with their flint-studded sickles to alter the genes. If they threshed the grain locally in the wild fields to get at the seeds, nothing special would happen. But if instead they took the cut stalks back to permanent villages, the agricultural revolution would begin.

Harris pointed out that the crude flint sickles shook the stalks of wild einkorn more violently than the small animals. Any spikelets with weak hinges would fall to the ground during the cutting. The strands that the gatherers bundled and returned to their homes would be variants with the strongest rachis. Its seeds would be larger. Unintentional spillage of these larger seeds around the village during threshing and grinding would produce crops like the variant, with the strongest rachis, and thus more productive and more desirable than those originally harvested farther away. Repeated cycles of cutting, spilling, and growing would select automatically for the genetic mutations leading to domestication. The inadvertently disturbed soil settings close to human settlements were to a certain degree the prepared seedbeds of garden plots. Those about to become the first farmers had merely to observe and connect cause with effect.

According to Harris no more than a few dozen generations of selective picking and accidental spillage would accomplish Childe's revolution. Ryan asserted that a drying lake might expose a new belt of moist coastline initially free of nature's cereals and grasses. If the foragers moved their villages to keep them close to the water supply, the spilled grain would find virgin soil at the lakeshore, and the step to domestication could be accelerated accordingly.

Harris did not agree that an oasis like the Black Sea lake was a potential

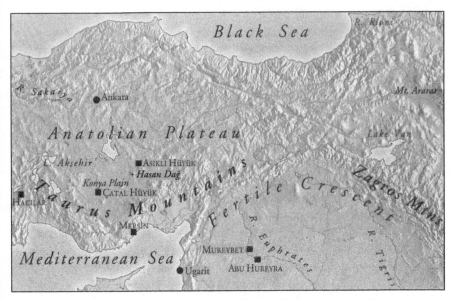

Anatolia at the time farming began to spread

cradle for the agricultural revolution. He argued that the cradle lay much farther south, where wild stands of the ancestral wheat and barley grow today on the hilly flanks of the Taurus and Zagros mountains. Domestication, he asserted, was a high-altitude phenomenon. Once accomplished, the genetically altered species might then be transplanted to lake margins and river floodplains at elevations near or below sea level.

DESPITE Harris's reservations, Ryan wasn't convinced that Pumpelly and Childe's oasis model should be discarded. In the spring of 1997, Ryan attended a symposium titled "The Late Quaternary in the Eastern Mediterranean," held in Ankara, Turkey, attended by archaeologists, climatologists, and geologists. They gathered at the headquarters of the Turkish Geological Survey for five days to present formal papers, and then, under the sponsorship of the British Institute of Archaeology at Ankara, traveled by bus for several more days to visit archaeological sites on the Anatolian Plateau, including the oldest farming town in Turkey. This was an opportunity to present the evidence of the Black Sea drawdown and abrupt flood to a knowledgeable and critical audience.

More than fifty lectures were presented. Those on the ancient lakes and vegetation history were especially informative concerning climate deterioration in the Younger Dryas. Ryan learned that there had been a number of very large lakes in Anatolia at the time of deglaciation. Lake Akşehir in the west was five times its present size as recently as the thirteen thousand years ago, but then in the Younger Dryas it had shriveled to a pond, leaving its beaches of cobbles and gravel high and dry. An enormous lake had occupied the Konya Plain in central Turkey until it also shrank under extreme aridity, its ancient shore smothered beneath large windswept sand dunes.

The most impressive documentation of Younger Dryas aridity came from Lake Van in eastern Turkey. Its surface fell by more than eight hundred feet in a few centuries. Equally startling was the evidence—from pollen and the ratio of the stable isotopes of oxygen in the shells of the zebra mussels—that the transition from dry to moist at the close of the Younger Dryas episode had taken ten to fifty years. Throughout the drought, the water in Lake Van and in the peripheral marshes had been far too salty for human or animal consumption. None of the Anatolian lakes could have served as oases. The Black and Caspian seas' lakes were the only possible candidates.

On the last day of the symposium the theme changed from seas, lakes, and landscapes to humans. Ofer Bar-Yosef from Harvard University gave the keynote paper entitled "Human Prehistory and Environmental Change in the Eastern Mediterranean." His initial slide engaged Ryan. Instead of displaying the first farming villages in the Near East in the typical arc of the Fertile Crescent—sweeping up through Israel, Lebanon, and Syria, and then heading eastward across Iraq before turning southeast into Iran (just as it had appeared countless times in textbooks and journal articles)—Bar-Yosef's version showed a large branch heading northwest across Turkey and almost reaching the Dardanelles and Bosporus Straits. Bar-Yosef introduced the Natufians and directly linked the impact of the Younger Dryas to the intentional cultivation of the wild cereals and grasses.

The rapid change from dry to wet at the close of the Younger Dryas led to the elevation of local water tables. The lakeshore and riverine valleys would have been just as assured of submergence from the switch of climate as they were from the opening of the Bosporus portal thirty-seven centuries later. To find the cradle of origins, archaeologists needed to look more broadly than where wild grain grows today. Putting himself in the archaeologist's shoes, Ryan asked Bar-Yosef, "Where would the Natufians have gone to find refuge given that they could not make it through the cold and arid spell in the Levant, even in the Negev and Sinai?"

The road map came in the very next lecture given by Nigel Goring-Morris from the Department of Prehistory of the Institute of Archaeology at Hebrew

University. Goring-Morris was an authority on patterns of population movement and was concerned with those caused by changing climate. He presented a strong correlation between climatic conditions and where people
chose to live. During the early deglaciation and warming of climate, people
had moved their campsites to progressively higher elevations in order to
position themselves near dependable water sources and live at comfortable
temperatures (atmospheric temperatures cool by 3.6 degrees Fahrenheit for
every one thousand feet of increasing altitude). These migrations had been
tracked in the Negev highlands. The wild grasses would have moved as
well and in warm times would have carpeted the flanks of Taurus-Zagros
Mountains, as they do today. But in the deteriorating conditions of the
Younger Dryas, humans, plants, and the game that follow the vegetation
would have descended back to the coastal plain or even down into the
Jordan Rift Valley, whose floor lies twelve hundred feet below sea level.

The Jordan Rift Valley was the focus of work by another renowned archaeologist, Kay Kenyon, who had served as director of the Institute of
Archaeology of the University of London between Childe and Harris. In the
early 1950s she had gone to Jordan to dig into an imposing oval edifice, Tell
es-Sultan, spanning ten acres of a rare oasis in the utterly barren landscape
of the Rift Valley floor. This mound sat next to a perennial spring, fed from
the Judean Hills to the west.

Kenyon had started her dig on the flank of the mound. As she moved to
the center through long, narrow trenches, she hoped to travel back through
time and confirm that Tell es-Sultan was indeed Jericho, the fortified Ca-
naanite city in the Bible that Joshua had ravaged forty years after the exodus
of Moses from Egypt. Kenyon fully expected that her incisions would bring
to light the very walls that had fallen down with the blast of Israelite trum-
pets. But to her dismay, no walls of the putative age of the return of the
Israelites to Canaan were there.

Kenyon persevered. She dug deeper, and one hundred feet down she
stumbled on the Natufians. The tiny flint flakes scattered all about her feet
displayed a precisely crafted design, each with a sharp straight edge and a
dull curved back manufactured in the shape of a half moon, used to make
the cutting blade of a sickle. Those at Jericho looked exactly like the so-
called tiny stone lunates previously discovered in caves on Mount Carmel.
Reaching down to pluck a lunate from near one of the postholes of a
Natufian hut, Kenyon spotted a harpoon head identical to those found in
the Mount Carmel caves. This artifact she knew to be characteristic of the
Natufian culture, which her carbon 14 dates would thirty years later place at
the onset of the Younger Dryas.

*A ten-thousand-year-old "plastered" skull exposed
in the wall of one of Kay Kenyon's trenches at Jericho*

As Moore did at Abu Hureyra, Kenyon found a farming village above the Natufian remains and separated from them by a sterile soil reflecting desertion there as well. The farmers at Jericho brought the same new house designs that Moore would discover in 1972 in Syria, the new tools, and the new form of burials: skulls removed from the corpse, covered with plaster, and given seashells for eyes, then placed in houses and shrines for veneration. Among the intricate stone blades of knives, daggers, and arrowheads of the Neolithic farmers, Kenyon recognized obsidian, a black shiny volcanic rock comprised of natural glass that cleaved like flint to make sharp, delicate blades. Years later this obsidian would be subject to chemical analysis, which would show that the first farmers carried this rock from volcanoes far to the north in eastern Anatolia near the headwaters of the Euphrates and only a few weeks' walk from the giant Black Sea Ice Age lake. Could this be a clue that the farmers came from the north and not from the south as generally accepted by specialists such as David Harris?

On the excursion after the meeting in Ankara, Neil Roberts took the participants to one of these volcanoes, called Hasan Dag. Neil had climbed to its snow-covered summit the previous summer and collected rhyolitic pumice characteristic of the type of material from the Mount Vesuvius eruption in A.D. 79. At the foot of Hasan Dag lay the ruins of a Neolithic village recently excavated by Turkish archaeologists from the Department of Prehistory of the University of Istanbul. The fifty-foot-high mound of Asikli sat on the north bank of the meandering Melendiz River at the place where it flowed through a narrow alluvial valley. The mound contained at least ten building levels. The walls from the deepest and oldest level penetrated could be viewed only through the gurgling water of the river in which it was submerged. These foundations date back to the centuries following the end of the Younger Dryas aridity. They had been flooded by the rising water table.

The layout of Asikli's houses, courtyards, and roadways pointed to advanced urban planning. Masonry consisted of both mud bricks and meticulously hewn blocks of Hasan Dag's volcanic tuffs. Beneath the plastered floors of rooms were burial pits—one containing the complete skeleton of a woman in her twenties with her infant. The mother had survived a "trepanation" in which a flap in her skull had been opened years before her death in a surgical procedure presumably to relieve pain or swelling. Necklaces of beads, semiprecious stones, and hot-worked native copper adorned her body, which was interred in a fetal position with knees drawn to the chest.

The excavation in 1989–90 exposed almost four hundred houses whose rooms varied from six feet on a side to almost fourteen feet. The hearths were either in a corner or against a wall, with a hole opened in the roof to serve as a chimney as well as an entry and exit. Mats woven of the cut stalks of grain or reeds had covered the floors. Neil Roberts pointed out the checkered imprint left by these rugs in the burnished plaster. Archaeologists found nearby Asikli and closer to the volcano the remains of open-air shops in which workmen knapped blades, scrapers, and knives from the obsidian to be used not just for themselves but for trade with other communities and for export to the Levant. Present in the scrap piles were fragments of polished mirrors still shiny in defiance of the long passage of time.

Roberts described how Asikli came to a sudden end during an eruption of Hasan Dag that buried the village and its tilled fields under a blanket of volcanic ash. He asked the tour group to stand in a semicircle on one of Asikli's streets and look over his shoulder at the summit of Hasan Dag. It appeared as two peaks. Pointing to the slightly higher cone on the right, Roberts said, "That is where I climbed to the crater and found signs of the last eruption."

Roberts then pulled from inside his jacket a piece of paper that he displayed for everyone to see. It was an enlargement of a photograph of the famous landscape painting from the north and east walls of shrine VII.14 at

Daily life in the stone-age farming town of Çatal Hüyük as it might have been during the millennium preceding the Black Sea flood

Çatal Hüyük, another Neolithic farming village eighty miles to the southwest and on the opposite side of the ancient Konya Lake. The nine-thousand-year-old painting showed the same twin-peaked panorama that filled the horizon before the group. Its Stone Age artist had drawn the taller right-hand peak in the midst of an explosion, volcanic rock flying upward in arching trajectories from the crown, a mushroom cloud of smoke high in the sky above, and incandescent flows of tuff rolling down the flank of the ten-thousand-foot edifice. The muralist had also depicted the planview of a village at the foot of the volcano, the floors of its rectangular houses laid out in a rigorous orthogonal street plan just like the one at Asikli.

"This is the remembrance of their demise," said Roberts. "The scene was painted maybe a thousand years after the event by an artist who could only have known of this history either from stories passed down through the ages or from other murals."

The town whose mural had kept alive the memory of the eruption was far better known to archaeologists and the public than Asikli. The discoverer of Çatal Hüyük was James Mellaart, who on a cold November night in 1958 had braved attack from local sheepdogs to reach the beckoning shape that had teased him for months on end as he excavated nearby. From his baptism as one of Kay Kenyon's fieldhands at Jericho, Mellaart knew the potential importance of this giant tell. Walking hurriedly around its base while there was still some twilight, Mellaart spotted stone tools all over the dusty soil. The scarcity of pottery immediately told him that the site was old. Walls of mud-bricks poked out of the side of the mound, and his trowel cut through a hump to reveal a surface coated with white plaster in remarkably good condition. Then from a coworker whom he had sent scrambling to the top he heard "Neolithic!"

"It had never been seen before," Mellaart explained to Ryan and Pitman in 1996. "Imagine, twenty meters [sixty-five feet] of Neolithic above the level of the plain! Right from the start I knew that not only was Çatal Hüyük old, but its occupation had lasted a long time."

Mellaart returned to Çatal Hüyük three years later to start a formal excavation. The thirty-two acres that he mapped with surveyor's instruments turned out to be three times the area of Kenyon's Jericho and ten times the size of Asikli. Upon opening up a small sector, like a single wedge of a layer cake, Mellaart uncovered the foundations of twelve building levels, each representing a new generation of construction. They spanned fifteen centuries.

Every house at Çatal Hüyük had its own walls. The dwellings were side by side without streets, not only for mutual support of adjacent walls but also to give access across adjoining roofs. Entrances and exits would have been through sky holes via a ladder. In nearly every room the built-in benches, platforms for cult objects, hearths, walls, floor, and ceiling were all coated with a fine white clay still in use today by Anatolian peasants. It was the plaster that Mellaart had noted during his first visit. The clay had been applied in multiple thin layers that the excavation team counted like tree rings to give a rough estimate of the lifetime of a domicile before it was knocked down, leveled, and used as the foundation of the successor house. Some walls had been repainted for more than a century, others for less than thirty years.

Mellaart relied heavily on carbon 14 to cross-check the relative ages of the different building levels and give him an absolute basis for correlation to Jericho. Most of his radiometric dates came from roof beams or wall posts in burnt buildings. Çatal Hüyük has divulged a history that Mellaart believes may well go as far back as Asikli's and Jericho's. Its lowest levels lie below the water table and can be opened properly only with pumps to drain the aquifer beneath the site. Beginning in 1995, after thirty years of inactivity, Çatal Hüyük was reopened. The new team is headed by Ian Hodder, who as a student had attended lectures by Mellaart at the University of London in 1967, the year of the publication of Mellaart's highly acclaimed book, *Çatal Hüyük—a Neolithic Town in Anatolia.*

Hodder had been greatly impressed by the mortuary practices and evidence of symbolism of these ancient people of apparently mixed races. Religion permeated daily life. The home served as a sacred space. In the darker recesses of the interior, Mellaart had found an elaborate and mystical ornamentation consisting of the skulls of the giant auroch (wild cattle) mounted on walls with a full spread of horns. He had also uncovered knobs that resembled the human breast with nipple erect, cracked open, and releasing a vulture. Empty eye sockets in jawless skulls of the village ancestors gazed out into the rooms from the tops of platforms. Creatures sculpted in bas-relief, such as a pair of facing leopards, adorned alcoves and recesses. The mother goddess representation greeted the observer with arms and legs outstretched from her protruding belly.

Although the plastered walls were commonly dressed in murals of natural landscapes, such as the eruption of Hasan Dag or scenes showing hunters in the pursuit of deer and auroch, some paintings displayed the macabre practice of leaving a recently deceased and beheaded corpse on a rack outside the town for its flesh to be removed by vultures. According to Mellaart's interpretation, this practice of "excarnation" was carried out not for reasons of hygiene but as part of a rite of passage, leading from death to an afterworld.

The sculpture at Çatal Hüyük depicted the bonds between the mother and the untamed world through the juxtaposition of women and leopards, in which the birth-giving mother cradles the feline cubs or sits beside or on them. The prolific Neolithic art included erotic imagery of animal heads emerging from the vulva between the outspread legs of faceless human bodies with bulbous bellies and full breasts.

For Hodder the painting and sculpture at Çatal Hüyük had direct parallels in the Iron Gate shrine of Lepenski Vir set high on the cliff overlooking the Danube River. Carbon 14 dating indicates that these sites in Asia and Europe were contemporary. The symbolism is ubiquitous from Palestine to Europe in the millenniums preceding the Black Sea flood.

NEIL Roberts took the participants of the symposium in Ankara to Çatal Hüyük expressly for the purpose of seeing the new clearings by Hodder and the careful attention being paid to conserve the decaying plaster walls with their colorful murals. It was here that Ryan could fully appreciate that Çatal Hüyük was really two tells, the larger, eastern mound of early Neolithic age and the smaller, western one made up primarily of later Neolithic and Copper Age levels. Mellaart had dated the highest level of the older eastern mound at 6200 B.C. while commenting that it had been deserted without signs of violence or destruction. A gap of hundreds of years separated those who quit Çatal Hüyük East from those who had laid the foundations of Çatal Hüyük West.

The gap reminded Ryan of the one observed by Andrew Moore at Abu Hureyra and the one found by Kay Kenyon at Jericho, although both were much farther back in time. The earlier hiatus had filled the second half of the Younger Dryas, when the climate had so deteriorated from aridity that the Natufians were forced to quit their villages. Now it seemed that the desertion of Çatal Hüyük, though much later, had been for a similar reason.

Scientific drilling of the ice cap of Greenland had returned a remarkable climatic record stretching from the present back more than 100,000 years. At a depth of 250 feet the packed and thermally insulated granular snow, known by mountaineers and glaciologists as "firn," turns to solid ice. At this transition samples of atmospheric gases become entombed in the ice in the form of tiny bubbles of trapped Arctic air. Since the polar ice sheet contains a thousand or more centuries of annual snow accumulation, its temporal record of precipitation and atmospheric composition is unequaled anyplace else on Earth. For example, the great chill of the Younger Dryas event stands out in remarkable detail in the polar ice record.

In the summer of 1992 a European consortium flew in 150 tons of equipment to the summit of Greenland at an elevation of more than ten thousand feet. Here they set up camp to drill a twenty-centimeter-diameter core of ice through the two-mile-thick ice cover. An American team joined them in 1993 for a companion hole. In addition to using the isotopes of oxygen to look at past temperature, the glaciologists were able to separate tiny bubbles of fossil air from the ice and measure its methane content. Methane (marsh gas) is a proxy of the activity of terrestrial biosphere (particularly the wetlands); its atmospheric concentration is elevated at times of warm, moist climate and reduced at times of cold, dry climate. In all the data, the methane measurements showed a second mini Ice Age in addition to the Younger Dryas. Counting the individual layers of frozen snow back in time, the polar scientists could date its beginning and duration. The cold snap started in 6200 B.C., coincident with Çatal Hüyük's desertion, and lasted almost four hundred years, until 5800 B.C. The amplitude of the methane drop was 90 percent of the value for the Younger Dryas, making it a second Mini Ice Age. Even Lake Victoria in tropical Africa responded to the polar cooling and drying by an abrupt drop in its water level.

Ryan pondered this younger cooling episode, and while touring the Çatal Hüyük excavation, he discussed it with Martine Rossignol-Strick, a palynologist who studies the preserved pollen in the organic residues of peat bogs, lakes, and oceans in order to reconstruct past climate and the light it sheds on the prehistoric environment. Rossignol-Strick had cut her teeth in Palestine piecing together the plant life around the Sea of Galilee in the Jordan Valley during the few millennia before the time of Christ.

Rossignol-Strick had recently put together a synthesis of all the pollen records from lakes and peat bogs in Greece, Turkey, Syria, Israel, Iraq, and Iran, and had established a chronology of climate change tied to the more reliable carbon 14 dates of seabed sediments. She asked Ryan if he had pollen data from the Black Sea. Ryan confirmed that both the Russians and the Bulgarians had carried out analyses, but since their cores were from the shallow shelf, there were no deposits in the time period of the young cold/ dry episode. The shelf was land then; the record there was missing. However, Rossignol-Strick was not discouraged. She said that she had professional contacts in Eastern Europe and would make inquiries.

In early October 1997 a large envelope arrived in Ryan's mailbox at Lamont-Doherty with pollen diagrams compiled by Bulgarian researchers and a copy of a personal letter from one of them to Rossignol-Strick. The letter commented that the transition from the Ice Age to the modern interglacial warm period had traditionally been recognized by the increase of woodlands and forest abundant in oak and a decrease of such steppe grasses as goosefoot and wormwood, and had never been independently dated in Bulgaria. An age of ten thousand years was given because this was when the oaks took hold elsewhere in Europe.

The pollen variations in the envelope came from Black Sea cores in water depths sufficiently deep so these sites were never above water in the partial desiccation. In these diagrams the appearance of oak, followed closely in succession by hazel, elm, beech, alder, and birch, was loud and clear. The changeover from cold and dry to warm and moist had not occurred on the Bulgarian coast ten thousand years ago as previously assumed. The radiometric ages in the Bulgarian cores showed the expansion of the forest starting only 7,500 years before the present. Before then vegetation had been unchanged since the onset of the Younger Dryas. The emerged shelf of the Black Sea prior to the flood was not a forest or woodland, as reconstructed by Andrew Sherratt, but was in fact grassland and steppe. A few willows and alder would have dotted the banks of perennial rivers flowing from the Balkans across the broad open coastal plain, but the oak, elm, beech, and birch had yet to establish a foothold. The setting would have been like the Konya plain in Anatolia in the days of Çatal Hüyük prior to its desertion. The Black Sea's lake was at an elevation three thousand feet below Çatal Hüyük. The regional cooling of an estimated 8 degrees Fahrenheit accompanying the newly recognized Mini Ice Age beginning in the late seventh millennium B.C. would have been more than compensated for by the difference in elevation between the giant lake and the Anatolian plateau. The Black Sea lake offered warmer temperatures and perennially flooded river valleys to cereals and grasses no longer able to survive the cooler temperatures in their previous habitat.

Like their Natufian forebears who were driven from Abu Hureyra and Jericho by Younger Dryas aridity and chill, the Neolithic farmers may have been forced from the Anatolian plateau by this second Mini Ice Age. In 1969, before Andrew Moore had opened Abu Hureyra, he had worked in Palestine where he found the era of farming there had also been cleaved by a desertion. Moore was impressed by the evidence he gleaned from site visits. Although reluctant to believe that Palestine had been almost entirely emptied for centuries, he wrote, "The archaeological evidence points to this conclusion." Within the resolution of existing carbon 14 dates from the Near East, Palestine's second desertion and that of Çatal Hüyük overlap and took place in the second Mini Ice Age that preceded the Black Sea flood.

Although Ryan and Pitman built their case for the Black Sea being Pumpelly and Childe's oasis entirely with circumstantial evidence and using facts gathered by others in disciplines far from geology, to the Columbia University scientists the Black Sea had all the prerequisites of an ideal refuge. Due to its setting below the level of the external ocean, it remained warm when the mountain flanks of the Fertile Crescent, the Negev highlands, and the Anatolian plateau chilled. It held vast volumes of fresh water when the lakes elsewhere shriveled to undrinkable salt ponds and marshes, and the Jericho spring dried up. Streams from the Balkans, the Alps, and Caucasus mountains kept the Black Sea's rivers in flow year-round when the Euphrates water no longer arrived at Abu Hureyra. The pollen from the bottom of the giant lake contained specimens of cereals and pulses, although it was as yet impossible to differentiate domesticates from wild progenitors from pollen alone.

Then in November 1997 the case for a northern oasis strengthened. Plant geneticists announced a match of the DNA in the earliest domesticated einkorn wheat (*Triticum monococcum* subspecies *monococcum*) from the earliest farming villages with a wild strain (*Triticum monococcum* subspecies *boeoticum*) living today in Anatolia. The fingerprint caught all the experts off guard; they had expected confirmation that the cradle of agriculture would turn up in the south.

None could have been aware of the second Mini Ice Age that was only now being discovered. Indeed, from all the evidence that surfaced at the symposium in Ankara and from the discussions afterward, it seemed quite likely that the humans who were there to witness the Black Sea flood and be driven from their homes by the inundation would have been townspeople, some skilled in tilling fields, planting seeds, harvesting crops, and breeding animals. They may even have been experimenting with the diversion of streams for rudimentary irrigation. Many would have been artisans—bricklayers, carpenters, painters, sculptors, basket weavers, leather workers, jewelers, potters, and morticians. Goods were made for both local consumption and for trade with other distant communities in the Levant and perhaps even in Eastern Europe as Gordon Childe had foreseen. A form of social and political structure would have been in existence, with one class of society conducting administrative tasks, others manual labor, and others such as the shaman performing ceremonies of religion, magic, and even brain surgery. They suffered from diseases including malaria and arthritis. The average human life span was barely thirty years, but a few elders lived into their sixties. One may presume that like their Natufian ancestors thousands of years earlier, when confronted by a drastic change in their environment, they would cope by packing their belongings and departing for a new homeland to carry on with their acquired knowledge, tools, and culture.

CHAPTER SEVENTEEN

The Diaspora

T HE ocean bursting through Bosporus in 5600 B.C. so violently cleaved Europe from Anatolia that it would have been several years before anyone dared make passage across. Likewise at the eastern end of the Black Sea, the snow-tipped Caucasus also presented a formidable barrier. Thus except for refugees willing to risk a voyage at sea, it is likely that those on the northern and western edge of the flooding Black Sea lake escaped into Europe and Ukraine, and those on the southerly side fled into Anatolia and points beyond.

Along the west and north shores several large rivers—Danube, Dniester, Bug, Dnieper, and Don—lead up through broad valleys deep into the interior of Europe and the steppes of Russia toward rich loess soils left by the Ice Age winds. These are likely arteries along which many refugees might have traveled not only to escape the drowning of their villages but to find new places to homestead. It is along such routes as these waterways that one might begin a search for evidence of migration.

In 1908 a Yugoslavian prehistorian named Miloje Vasic recognized the remains of ancient human settlements poking out of the eastern bank of the Danube River about ten miles downstream from Belgrade. The strata, peppered with bones, stone tools, and broken pottery, was nearly thirty feet thick. In it he found two superimposed cultures. Over a period of twenty years he sorted through the layers, showing off his various treasures to visiting archaeologists and asking for help in placing them in a context consistent with artifacts surfacing elsewhere in Romania and Hungary. He put many of the prized objects on display in a small museum so that the local citizens could view their heritage.

One of the peoples discovered by Miloje Vasic in 1908, the Vinča, occupied

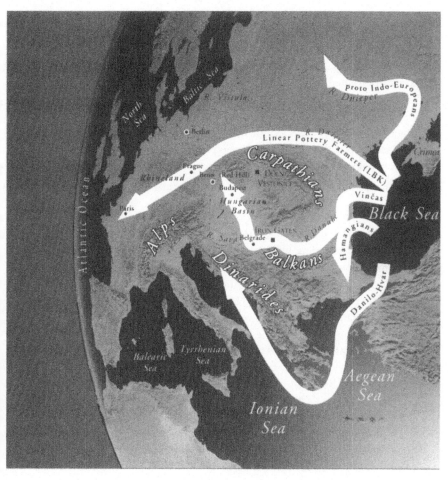

*Inferred human migrations west and northwest into Europe
in the wake of the Black Sea flood*

the site on a terrace above the Danube visited by Childe and his companions in 1926. It seemed clear to Vasic that the Vinča had built on the deserted ruins of an older culture. Makers of lovely wattle and daub houses and fine incised pottery, the Vinča appeared abruptly on the plains of Bulgaria within a century and a half after the flood, settling also on the river terraces of the southern Hungarian plain and in mountain valleys as far south as the Vardar River in Macedonia. They constructed well-planned permanent villages on leveled ground with parallel rows of houses separated by streets. Unlike their predecessors, they built villages one on top of the other, staying on one site for successive generations. They plastered their floors with white clay. But instead of constructing their walls of mud-brick, they built them from split timber planks or hewn posts interwoven with twigs and covered with a thick layer of mud plaster. Archaeologists have uncovered shrines decorated by bucrania, attached to a wall beam as in the shrines of Anatolia.

Vasic saw no continuity between the Vinča culture and the underlying strata of their predecessors but rather thought the Vinča were outsiders who settled on a previously abandoned site. Their art and pottery were so exceptional and in such contrast to the prior occupants that Vasic mistakenly identified this "as a center of Aegean civilization in the second millennium B.C."

Curiously, all the Vinča settlements are located well back from the sea. Only in the European interior, on the Hungarian plain, within the protective ring of the surrounding mountains, did they live on landscapes below an elevation of three hundred feet. Could it be that they had experienced the catastrophe of the flood and feared the sea?

Another group of farmers called Linearbandkeramik (LBK), meaning "linear-band ceramic," a name derived from their distinctive style of pottery, appeared in Europe at the same time as the Vinča, rapidly occupying an arc from the Dniester River across northern Europe as far west as the Paris basin, displacing the indigenous hunter-gatherers. To a number of experts the spread of the LBK culture along this arc reflects colonization by farming populations in such a brief period of time that its beginning and end are at present unresolvable by the radiocarbon dating methods. They brought with them their longhouse building style, never before seen in Europe; these huge timber-framed houses, up to 150 feet in length, were organized into villages founded exclusively on the fertile loess soil blown across Eurasia during the sky-darkening sandstorms of the last Ice Age. These dwellings were the largest freestanding buildings in the world for thousands of years. This departure from the earlier mud-brick house of Anatolia and Greece may reflect the availability of wood and their dispersal away from the dry microclimate in the Black Sea depression.

As the name suggests, the LBK pottery was decorated almost exclusively with incised patterns of parallel grooves and bands of dotted lines forming spirals, waves, concentric rectangles, and other geometric designs, almost all without applied color. The absence of pigment may reflect an original homeland, such as the incised river valleys out at the edge of the Black Sea lake, where the minerals for making the color paints were nonexistent.

A very striking feature of the LBK is the homogeneity in pottery design, stone tools, village plan, house shape, burial practices, and economy over the vast territory into which these people appeared, suggesting that their dispersal was almost instantaneous. Experts specializing in pottery from Belgium can readily recognize shards from Moldavia as if they had been crafted nearby in France. The domesticated plants and animals show practically no variation from village to village across a span of a thousand miles or more. But there is a dramatic cultural gap between the preexisting sparse hunter-gatherer population and the LBK homesteaders. The new arrivals either absorbed those in their way or wiped them out. LBK fortifications occur along the periphery of their final expansion and are separated by a no-man's-land with no sign of coexisting foragers.

Their explosive movement from east to west up the Dniester and Vistula rivers and across the Rhineland to the valley of the Seine has only recently been recognized as a mass immigration, almost an invasion.

Like the Vinča, the Linearbandkeramik never put down permanent roots near a sea coast. They never colonized fertile land in the coastal regions of northern Europe. Neither did they settle along the postflood coast of the Black Sea, stretching from Turkey through Bulgaria, Romania, Moldavia, Ukraine, and Russia. Was this a response of people who had been chased upstream by the flood and who still worried that the fountains of the deep would spring forth again?

Simultaneously with the appearance of the Vinča and the LBK, in the mid sixth millennium B.C., the Danilo-Hvar settled on old abandoned sites along the Adriatic coast of Dalmatia in several of the fertile valleys that cut through the mountains to the sea. Marija Gimbutas proposed that they had come from somewhere in the southeast. They, too, stayed well inland of the coastline. Strategically located at the seaward end of the Neretva River valley, the Danilo-Hvar provided a trading link between the Adriatic and the mountains and valleys to the east via a closely related group of newcomers called Butmir who occupied the lands around Sarajevo. They exported *spondylus* shells across the Balkans to the Black Sea side and imported obsidian from Italy. They may have purchased salt from as far away as southern Poland. Danilo-Hvar pottery was sophisticated and beautifully decorated with rich patterns of chevrons, spirals, running waves, nested S's, and other geometric figures, sometimes painted in black and red on white slip but at other times incised. They crafted a now-famous pot decorated with a sailing ship, depicting masts and rigging dated at about 4000 B.C.

People called Hamangians also seemed to emerge out of nowhere to settle in the region of coastal Bulgaria. Two fascinating and quite modern-looking sculptures from early in the fifth millennium B.C. were found together in a grave. One dubbed "The Thinker" is the figure of a man seated on a low stool, legs bent, elbows on his knees, hands on his cheeks. The other is of a woman seated on the ground, one leg stretched out in front of her, the other bent and upright on which she is resting her hands. Both portray a state of quiet contemplation and repose, in striking contrast to the styles of their predecessors. When the Hamangians were first discovered, it was suggested, on the basis of their splendid sculptings, their use of marble, and the presence of *spondylus* shells, that they were immigrants from the Levant or somewhere else in southwest Asia. Could they have come from the area of Crimea or even the southern edge of the Black Sea lake perhaps making the journey by boat? These may be the only group who seem to have been willing to settle near the sea.

All these people appeared in Europe shortly after the flood. All have been described as outsiders: people who migrated from some distance, although this is a point of contention for some archaeologists. All seem to have been more culturally advanced than those whom they replaced. Perhaps not so coincidentally, at that time in the middle of the sixth millennium B.C., Europe began a rapid ascent into what Childe and Gimbutas have called a "Golden Age," a transition that in Gimbutas's words "has led many scholars in the past to assume that tides of colonization must have burst through the Balkans. They would have come, it was thought, either from Anatolia or from the eastern Mediterranean." Or, as we would suggest, from the Black Sea rim, an influx of people with new ideas and crafts, giving a fresh impetus to what was already there and lifting southeast Europe over the next several hundred years to new cultural heights.

Trade flourished as never before, perhaps ignited by the abrupt influx of displaced people who had lived by the edge of a vast lake and who had probably been sailing and trading along its shore for many generations. This broad exchange of wares stimulated the manufacture of goods to be traded, in particular those crafted in copper, which further invigorated the barter economy. Axes, beads, rings, arm bands, and pendants made of this pliable metal have been found at sites ranging from the rivers feeding the Adriatic to those pouring into the Caspian Sea.

It may have been nothing more than the forced exodus of more advanced peoples from a grand melting pot in the wake of a Great Flood that gave the culture of Old Europe its thrust to a "Golden Age."

An exodus of Black Sea folk to the south would have been an entirely different enterprise because the exposed rim of the Black Sea lake would have been considerably narrower along its southern and eastern coasts. This southerly rim abutted a mountainous terrain, whose rivers were mostly fast flowing and unnavigable for any significant distance. The river arteries would have led to a drainage divide in the open and dry Anatolian plateau. Furthermore, this plateau had its own population which had settled in an area that due to the climate stress of the younger Mini Ice Age may have been populated to the sustainable limit. Also, to the south and east the climatic effects of this second cooling and drying seem to have been more severe. Many villages in Anatolia and the Levant had been deserted.

But this period of cool aridity ended about two hundred years before the flood, so that as the climate began to improve and the droughty conditions of the Levant, northern Mesopotamia, and Anatolia eased a few small groups of migrants may have wandered out of the Black Sea basin to occupy some of the previously abandoned sites of this newly refreshed landscape.

Early in their search for a sign of the Black Sea flood in the archaeological record, when the deluge itself seemed a sure thing but its effect on humankind was an open question, Bill Ryan and Walter Pitman discussed the problem with the distinguished archaeologist Ralph Solecki. This emeritus professor at Columbia University had extensive training in geology and was knowledgeable about glacial and postglacial landscapes and climate. He had explored caves and riverbeds in Turkey and Iran in search of the remains of early man.

Solecki explained that the archaeologist would look for the sudden building of walls around villages as a sign of mass population movements. Fortifications would be erected when resources were limited and people felt threatened by neighbors or outsiders. During times of peace the walls would be neglected and allowed to decay. He noted that evidence of hostility is conspicuously lacking in the Neolithic record from southwest Asia. His colleagues digging in Iraq, Iran, Syria, and Jordan had reported signs of peaceful cohabitation. They had uncovered villages without walls and with no defensive weaponry. Few of the hundreds of skeletons exhumed from the graves possessed wounds attributed to warfare. One puzzling anomaly was the great stone wall discovered by Kay Kenyon in the lowest levels of Jericho; it dated to the tenth millennium B.C., and some experts believe it was erected for flood control rather than for defense.

By the middle of the sixth millennium B.C. exceptions appear. In the mound of Hacılar in western Anatolia, part of a fortress had been unearthed

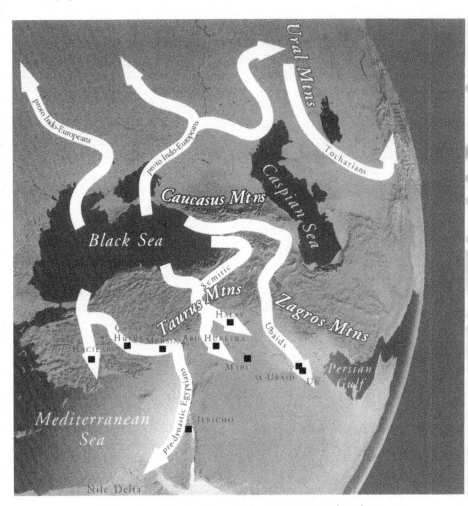

*Inferred human migrations northeast into Asia and southeast
into the Levant, Egypt, and Mesopotamia*

by James Mellaart in 1957. It belonged to the terminal sequence of a village
whose foundations were established fifteen hundred years before the Black
Sea flood. Like the main town of Çatal Hüyük, Hacılar also had a gap in
occupation during the cold climate of the late seventh millennium B.C., only
to be resettled and then continuously occupied a few centuries later. The
craftsmen who reoccupied Hacılar after its desertion created sophisticated
painted pottery, more technologically advanced in style and fabrication than
any contemporary pottery found elsewhere in the entire Near East. Fantastic

curvilinear figures painted in red on a cream background attest to a recently learned technique that could have been learned at the site from whence they came, perhaps from others who took the same refuge from the cold and dry climate.

At Hacılar, Mellaart had uncovered five building horizons in the upper strata above the gap. They revealed a complete succession of cultural continuity until a disturbance found practically at the surface of the mound. For the first time fortifications emerged in the digging. They occurred right below the disturbance and comprised a section of the village cut off from the rest of the community and encircled by a defensive wall. These ramparts enclosed not only houses but a large granary for storage, a well, a potter's workshop, and a shrine. That section became a citadel for siege against an enemy. The potter's kiln would have been utilized to make hard clay balls that were flung from the ramparts by sling.

The ramparts were destroyed shortly after completion, and the invaders built on its ruin a new fortress. Mellaart described his astonishing find: "Considerable changes in painted pottery . . . suggest newcomers with different techniques and traditions." From Mellaart's carbon 14 dates it seems that Hacılar was conquered close to the time of the abrupt flooding of the Black Sea. It is possible that these newcomers were people from the north who had been driven out by the flood and were in desperate need of food and shelter.

The patchwork of coastal plains, arid steppe, and rich valleys in the Levant and the northern Fertile Crescent have sustained humans for thousands of years. During the time of the late seventh-millennium B.C. Mini Ice Age, many of these valleys and their villages were deserted.

The carbon 14 age determinations obtained from sites in the Levant do not have the resolution needed to confidently relate the desertions and reoccupations precisely to these climatic events. That there was at least one long period of abandonment, probably due to a climate crisis, seems certain. But in spite of the imprecision of the data, one can safely say that, at about the time of the Flood, a number of deserted villages were reoccupied and new settlements appeared, particularly along the coastal plains of the Levant. People called Halaf, after the Tell at which they were discovered, seemed to make their first spotty appearance in the fringes bordering northern Mesopotamia, in the northern Levant, and in Anatolia early in the sixth millennium B.C., at the end of the second Mini Ice Age. But again the timing of this event is uncertain. The Halaf were farmers and herders who made beautiful pottery that for the most part was decorated with exquisitely detailed geometric designs. Some of their buildings were dome shaped, like a beehive. They

*The elaborately decorated pottery that appeared in Syria and Mesopotamia
in the centuries bracketing the Black Sea flood*

were built entirely of dried mud and mud-brick; they may have been more
or less impervious to rats and other vermin, and thus used principally for the
storage of foodstuffs. Some insist that the Halaf emerged from the indigenous
population. Others have assigned a northern or Anatolian origin to the Halaf.
They could be among those who began to wander out from the margin of
the Black Sea lake when the effects of the cold, dry climatic event had
ameliorated. As Moore has pointed out, there was a sizable influx of farming
peoples along the coast of Lebanon and in its valleys in the mid sixth
millennium B.C. Were these refugees from the flood?

Egypt had experienced a rapid cultural and economic change during the
same period, at the time of the flood. A new flint industry was introduced,
epitomized by two-sided flaked tools, which was much more in common
with the industry at Çatal Hüyük, Hacılar, and Jericho than with the preceding African designs. In addition, the art of pottery making appeared for the
first time in the Nile Valley. Domesticated cereals and animals with direct
genetic affinity to Asia were also suddenly adopted, along with the first
systematic practice of planting and harvesting in fields watered from the
Nile.

In recent years archaeologists have reported the sudden appearance of advanced farmers along the Rioni River in Transcaucasia, midway between the Black and Caspian seas. With no precedent, no roots, and seemingly without forebears, they built a town of mud-brick buildings (some were dome shaped like those of the Halaf) and planted fields with grains and pulses. They may have used a crude plow and dug ditches for simple irrigation. The overlay on an earlier prefarming community was so sudden that archaeologists view it as an immigration from another part of the Near East. Carbon 14 dating places the settling of the Transcausus contemporaneously with the beginning of the LBK dispersal, the defense and fall of Hacılar, the arrival of newcomers in the Levant—particularly in the valleys and along the coast of Lebanon and at Tell Ard Tlaili in Palestine—the introduction of Asian domesticates in Egypt, and the flooding of the Black Sea.

To the east of the great Syro-Arabian Desert lie the fabled lands of Mesopotamia. Contained within and between the valleys of the Tigris and Euphrates rivers, this arid near desert is said to have been the cradle from which western civilization sprang. It is a land of extremes. A gently rolling landscape of shallow valleys and low-lying smooth hills with deep wadis in the north gives way in the south to a flat and intensely hot alluvial plain. The extreme aridity is relieved only by the flow of the two great rivers. Within this unforgiving scorched landscape of southern Mesopotamia, and thanks to its immensely fertile alluvium, the first great city-states arose.

From headwaters deep in Anatolia, the Tigris and Euphrates flow in their own broad valleys that are confined to channels between rich fertile terraces. Water for irrigation must be lifted from the river to nurture the lush gardens, orchards, and fields of crops. Outside the valley the tableland is a near desert. To the south the valleys of the two rivers merge into one broad alluvial plain across which the rivers meander. Much of the terrain has been made inhospitable by thousands of years of overuse and the gradual incursion of salt water. Rows of poplar trees border the waterways. Clumps of tamarisk bushes sprinkle the landscape.

The annual rainfall on the alluvial plain is less than ten inches per year in the north and as little as three in the south. Farming is possible only with irrigation. Except for a few isolated settlements where its margin meets with the foothills of the Zagros Mountains, the land between the great rivers south of Baghdad was unoccupied at the time of the flood. The people who arrived later would turn the land into a breadbasket. Known as the Sumerians, they went on to create one of the most impressive civilizations the world has known.

The Sumerians were thought to be descendants from a distant homeland to the north. In searching for a linguistic affinity with modern dialects, Henry Creswicke Rawlinson reckoned that their particular use of pronouns was more like the languages of Mongolia and Manchu than any other type of Asian family. One of his most respected colleagues pointed out what he believed were close ties with Turkish, Finnish, and Hungarian. A modern Assyriologist, J. Bottéro, writing in 1987 on the literature, reasoning, and gods of the Sumerians, stated that "we do not know anything of their earlier ties, as they seem to have burned all bridges with their country of origin, from which they never received any new blood, as far as we know."

The Sumerian view of their own alien past is expressed in their "Poem of the Supersage" in which the Great Flood marked the end of mythological time and the inauguration of historical time. It is intriguing that they believed the seven sages appeared from the sea during the "first days" in human form wearing fish skins. In the epic of *Gilgamesh* the seven sages are credited with building the walls of Uruk and bringing the arts of civilization to the Sumerians—irrigation, farming, and the use of copper, gold, and silver. The question of where the Sumerians came from is still unanswered.

During the several millennia before the flood, farmers had settled along the foothills of the Taurus and Zagros mountains, where winter rains and springs fed the numerous rivers and streams, and provided moisture for a more verdant landscape. However, none of these people possessed the technical skills to challenge the southern alluvium between the two great rivers. On the floodplain the basic resources were two: the reliable and constant flow of the rivers and the rich fertile soil. The challenges were many and formidable. Stone for making tools and wood for building, cooking, and firing kilns were completely lacking. Temperatures rose to over 120 degrees Fahrenheit in the shade. Only a few inches of rain fell each year. Unlike the Nile Valley, the annual floods from the melting snows of the mountains were irregular and late in the planting season. Irrigation was not an option, it was a necessity.

It was up to Charles Leonard Woolley to make the first serious excavations on the alluvium of southern Mesopotamia and reveal the extraordinary talents of the Ubaids, its first farmers. In the era between the two world wars, he opened the Tell al-Muqayyar, meaning "the mound of pitch," known as Ur to the ancients. The whole desert north of the tidal swamps at the head of the Persian Gulf is dotted with mounds. Out of this scorching land rises Ur of the Chaldees, the biblical home of Abraham and the highest edifice of them all. While excavating at Ur, Woolley spotted, across the monotonously flat landscape, the shadow cast by another, yet undiscovered tell of great historical importance.

Woolley wrote:

Standing on the summit of this mound one can distinguish . . . to the north-west a shadow thrown by the low sun may tell the whereabouts of the low mound of al-Ubaid.

No previous visitor had ever noted the shadow of al-Ubaid. For had a caravan passed and stopped for rest, the traveler would have stumbled on thousands of fragments of painted pottery, flint, chert, obsidian, carnelian, crystal flakes, disk-heads, pegs, pieces of aragonite vases, inlay plaques of red stone, and copper nails which strewed the desert with the relics from the first successful colonizers of the desert alluvium. Its conquest was made possible with a new farming technology based on irrigation. Its increasing sophistication became the prime mover of a growing economy. The challenge of irrigation was as much one of control of water as acquisition. One needed to get water on the land, hold it there for as long as needed, dispose of it when no longer required, and keep unwanted water away.

The Ubaids built large houses for their extended families. At the level of earliest occupation three homes have been excavated at Tell el Oueili, just across the Euphrates from al-Ubaid. One of the foundation dwellings was over forty feet in length with an external staircase flanking a hall whose roof was supported by rows of posts or columns. Wood, which had to have been imported from far to the north or east, was from the beginning an important building material. The obsidian for their serrated knives came from Anatolia. Their terra-cotta figurines reveal tattooed bodies—the women strikingly slim with bands around the neck and waist. Within the uncertainties of the carbon 14 dating methods the settlement of al-Ubaid appears to date to the era of the arrival of newcomers in Palestine and to the Black Sea flood.

Many of the Sumerian cities, such as Eridu, sit on Ubaid predecessors without any sign of occupational interruption. The massive temple at Eridu was the home of Enki, the god of the freshwater ocean, whose cult worship can be traced to a meager hut at the time of the Ubaid colonization of the southern Tigris-Euphrates floodplain. Woolley is famous for his discovery of the royal cemetery at Ur where the queen and her whole entourage of servants and guards were entombed in a "death pit," a time capsule of 3000 B.C. Her retainers were sacrificed along with the oxen pulling her possessions on carts to the netherworld. Sir Arthur Keith describes the skeletal remains in Woolley's expedition report:

> The southern Mesopotamians . . . had big, long and narrow heads. . . . Their affinities [were] with the peoples of the Caucasian or European type. . . . We may regard south-western Asia as their cradleland until evidence leading to a different conclusion comes to light. They were akin to the pre-dynastic people of Egypt.

In the Sumerian language most of the root words are monosyllabic. However, those having to do with agriculture and crafts are polysyllabic, such as the words for farmer, herdsman, fisherman, plow, furrow, metalworker, blacksmith, carpenter, basketmaker, weaver, leatherworker, potter, mason, and even merchant. These may have indeed been brought to Mesopotamia from the Black Sea melting pot on the journey south and later passed from the Ubaids to their Sumerian successors.

The oldest known written versions of the flood were committed to clay tablets over two millennia after the event in Sumerian, the language of the first known writing, a language with no known roots and no known descendants, and Akkadian, one of the ancient tongues of the Semitic language group to which the Arabic dialects and Hebrew belong. Is it possible through linguistics to tie these people together, with speakers of other languages at about the time of the flood, and to the region of the Black Sea.

Linguists have discovered that languages evolve in ways that are analogous to biological species. Linguists construct tree diagrams, showing the development of families of languages similar in design to those used by biologists to illustrate the path of evolution of humans and their primate cousins. It is certain that all of the languages spoken by the various people driven out by the flood have withered, just as Latin is no longer a spoken language. Traces of these ancient languages might be found in the ancient writings and the modern languages that exist today in Europe and southwest Asia, the territory to which these people fled at the time of the flood, just as traces of Latin may be found in many European languages. Linguists have developed techniques to reconstruct elements of languages far past their earliest written form. Times when people speaking different languages occupied the same or adjacent territory may be detected. In addition it may be possible to attach these ancient people to the time when their language was written, when their myths, beliefs, and perceived histories were finally committed to writing.

Genetic studies may provide another window in time. Unless exterminated by some disastrous event, those who fled would have continued to propogate, leaving their genetic traces to be passed along through the vast human gene pool. From studies of the present-day distribution of genetic materials, it is possible to look into the past and make intelligent appraisals of ancient dispersions—who came from where and when. The deeper into the past one looks, the vaguer and foggier the picture.

CHAPTER EIGHTEEN

Family Trees

"It was a beautifully preserved roll of paper, about a foot high and perhaps fifteen yards long, which I unfolded with Chiang in front of the original hiding place," explains Mark Aurel Stein in his 1912 personal narrative *Ruins of Desert Cathay*. "The writing was indeed Chinese; but my learned secretary acknowledged that to him the characters conveyed no sense whatever."

Astonished, Stein and fellow explorer Yin Ma Chiang scanned the scroll from beginning to end in the dusty grotto carved from rock. They were dumbfounded. All that they could decipher from their close examination of the writing was a frequently repeated formula. Perhaps it was a chant in a non-Chinese language whose phonetic sound was expressed by the ideographs. With Chiang's help Stein tried pronouncing the two side-by-side glyphs that had repeatedly caught his attention as the parchment was scrutinized under the guard of the temple priest. "Pu-sa," Stein uttered. The priest nodded in recognition. "Pu-sa," Stein repeated. The priest placed the palms of his hands together and bowed his head, unaware that this word, so familiar in his prayers, was the ancient Sanskrit term "Bodhisattva," dating from the first century A.D. when Buddhism had first entered the Tarim basin in Chinese Turkestan.

A few days earlier, vague rumors had surfaced that hinted of a hoard of ancient manuscripts stashed away in the caves of the "Thousand Buddhas," adjacent to Stein's campsite. The Hungarian-born adventurer had trekked three thousand miles on foot and entered central Asia from India on a tortuous path up the Oxus River to its cradle in the snowy ranges of the Pamirs. Threading the Mingtepe Pass, he had descended from the "Roof of the World" through the gorges of torrential glacial streams that flowed toward the Takla Makan Desert of western China. Most of the watercourses cascading down the margins of this intermontane depression never reached a lake. Instead, they withered from evaporation and dead-ended in sand dunes. Stein found himself at times retracing the track Marco Polo had taken during his journey to Cathay seven centuries before. The trail led to a rare oasis that long ago had become a sort of Mecca for Buddhism.

Religious piety had honeycombed a local cliff face into hundreds of cave temples, many richly decorated with frescoes and stucco sculpture, and some still venerated by Taoist monks who tended the ruins. In one of the hand-hewn vaults, during the spring of 1907, Stein gained access to the treasure for which he is remembered—a great deposit of ancient manuscripts and art relics that had lain hidden and perfectly protected in a walled-up rock chapel for about nine hundred years. The assertion that some of these manuscripts were not Chinese made him eager to learn the source of the unknown language.

Some of the manuscripts were bilingual. With the aid of the Sanskrit rendition, the extinct language surrendered to comprehension. It was given the name Tocharian after an ancient Persian tribe described by the classical writer Strabo. Tocharian displays conspicuous similarities in vocabulary and grammar to Celtic and German. These attributes identify Tocharian as a limb on the trunk of the Indo-European language tree, one of the world's major language groups, descendants of which are spoken today by people in Eurasia as far ranging as India, Iran, Iraq, Pakistan, Afghanistan, Armenia, almost all of Europe, and North and South America.

The Takla Makan Desert had preserved not only the fine parchment and palm leaves with their Tocharian writings but also corpses of the former inhabitants of the region in which the language was spoken. These people are far more ancient than the manuscripts. A naturally mummified woman, her desiccated flesh in an incredible state of preservation, was exhumed from the desert floor of the Tarim Basin by a crew of Chinese archaeologists in 1985. The mummy was dressed in a colorful blouse, skirt, leggings, boots, hat, and jewelry. She is estimated to have been fifty-eight years old and to have lived at least a thousand years before Christ. What most surprised her discoverers was her almost six-foot height, her long, narrow nose, her thin lips, her deep-set eyes, her blond hair, and her pointed skull, all unmistakable features of a European pedigree. Some graves contained woven plaid textiles. The diagonal twill weave requires a type of loom never before linked with central Asia. A specialist in textile archaeology at the University

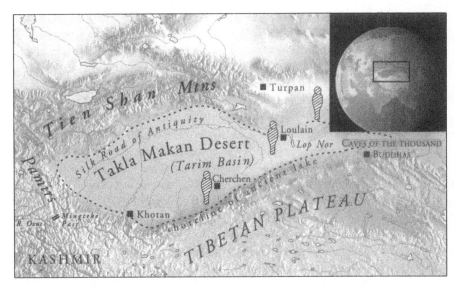

*The Takla Makan desert of western China with the shoreline
of the giant lake that formerly filled the Tarim Basin*

of Pennsylvania remarked that the wool design was "virtually identical stylistically and technically" to plaids found in Germany and Austria from roughly the same time.

The perplexing discoveries of a lost language with European connections and then actual bodies with unmistakable Caucasian features so far away as Chinese Turkestan was intriguing. Who were these people? From where did they arrive? When did they make their journey? Why had they chosen such an out-of-the-way place to inhabit?

SUCH questions necessitate detective work able to reach beyond the boundaries of the traditional disciplines of prehistory. One relatively new field of exceptional promise is human genetics. Its contribution begins in the former capital of the Lombard and Carolingian kingdoms of Italy in the mid-twentieth century with a professor at Padua University asking, "What makes individuals and populations biologically different?"

Early in his career Luigi Luca Cavalli-Sforza began to explore the role that chance plays in the hereditary process. What causal mechanism leads to the spreading of a mutation? Chance, or more properly genetic drift, he learned, is especially important in small populations. Genetic drift takes place because the genes of individuals sometimes undergo spontaneous change or mutations. These are usually quite minor and have no noticeable effects, and are often eliminated from the gene pool in one or two breeding cycles, but some survive. The progressive effect of a number of these mutations on a population is that the genetic makeup of the population gradually changes. This process affects all communities at rates that depend on the size and degree of isolation of the community. In large, genetically diverse populations, such as in a city with a constant flux of people moving in and out, the effects of genetic drift or random mutation tend to be diluted, and as a consequence the rate of genetic change is slowed. In small, isolated villages in which population exchange with the outside world is infrequent, these spontaneous genetic changes will soon spread throughout the village and its immediate surroundings. In this case, undiluted by a large and changing gene pool, the passing on of genetic change can be quite rapid.

Cavalli-Sforza decided to try to quantify the process of change through chance. He chose the valley of the Parma River in the Emilia-Romagna province of Italy as a study area. With the help of parish priests, Cavalli-Sforza and a colleague were able to take a significant number of blood samples from the local population. The two scientists filled their syringes in large prosperous towns on the alluvial plain, then in middle-sized villages upriver in the foothills of the Apennines, and finally in isolated hamlets high in the mountains containing but a dozen or so families. They were equipped to investigate only a few genes such as give rise to the A, B, and O blood groups and the Rh negative and positive antigens. The blood was drawn in parish sacristies after Sunday mass.

"Our findings fully confirmed our expectations, sometimes right down to the smallest detail," Cavalli-Sforza reported forty years later in a book coauthored with his son. Without exception the greatest variation, reflecting a rapid spread of individual genetic mutation, was found among the most remote hamlets; in the large towns there was the least variation because random genetic fluctuation in towns was overwhelmed by genetic variety. The magnitude of this so-called genetic drift depends on the number of individuals in the population and the degree of breeding with outsiders. For his Parma River study, Cavalli-Sforza used the parish records for the whole valley, recording births, deaths, and marriages in each village back to the sixteenth century. He mused, "It may seem strange that the study of the effects of chance can be predicted." But that is exactly what he had been able to do.

Most groups do not stay isolated from each other over long periods of time except in unusual circumstances. Migration generally takes place on a

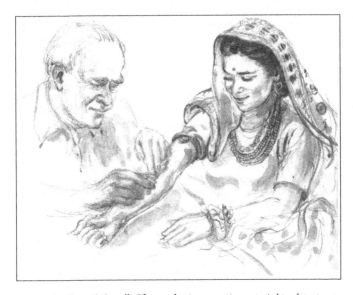

*Luigi "Luca" Cavalli-Sforza obtains genetic material to detect
and map the human diaspora*

small scale between neighboring villages. As a consequence, villages close by are more similar in their genetic composition to one another than to distant ones. Far-ranging migration of an entire group is possible but rare. Such diaspora happen in the wake of a great famine, war, or natural disaster. When a group traveled across a continent or between continents in ancient times, all contact might be severed between the immigrants and their original homeland.

As Cavalli-Sforza was arriving at the conclusion that mass migration leads to greater genetic diversification among groups, he discovered additional support in the data of a hematologist from northern Spain. Michael Angelo Etcheverry had also been drawing blood. His study focused on peasants, shepherds, and fishermen in the Basque provinces of the Pyrenees Mountains and in coastal cities facing the Bay of Biscay. The Basque people have their own language, unrelated to any other in Europe or elsewhere in the world. The Basques had fought off assimilation by the Romans, Visigoths, Moors, and Franks. Etcheverry had found an anomalously high frequency of the Rh negative antigen in the Basque population as a whole, almost two times the level in the rest of Europe. In a brilliant deduction he suggested that the living Basques were direct descendants of Ice Age Cro-Magnons who left the signs of their remarkable hunting culture in cave paintings scattered throughout the Iberian peninsula but especially concentrated in Basque country. The indigenous foragers south of the great mountainous divide of the Pyrenees had apparently avoided to a much greater degree than others north of the natural barrier the modification of their genetic heritage during the expansion of Neolithic farming commencing in the mid-sixth millennium B.C.

Cavalli-Sforza then speculated that "if enough data on a number of different genes are gathered, we might eventually be able to reconstruct a history of the entire human species." Perhaps the field of genetics could erect its own evolutionary tree, he thought. By 1984, the tools of heredity had advanced from a simple blood type to the classification of molecular DNA and up to nearly a hundred different genes. The increasing number of variables that could be measured from each of the hundreds of blood samples taken required a method of computational mathematics for the analysis. Statisticians and applied mathematicians were recruited to create equations and computer code. Scores of individuals from distinct native populations were typed for their genetic makeup. Out of the rows and columns of the lengthy printout, a set of maps of the genetic landscape emerged, each portraying an independent piece of information formerly hidden in the complexity of the raw data.

According to Cavalli-Sforza, the genetic data appeared to reflect the migration of a farming people from Palestine and Anatolia into and across Europe, meeting a local population of foragers en route, interbreeding, and continuing the journey one generation after another. The picture seemed to show a wave of advance from east to west. The ancestors of the Basques had lived at the far end of the migratory path. That explained why they had undergone the least genetic admixture with the newcomers—that and the fact that they must have hidden in their mountain refuge and avoided contact. This also explains why their language shows no relationship to the other European languages.

Cavalli-Sforza was ecstatic. He had found a key to the past. The modern geography of human genes was simply the current snapshot in the evolution of intermixing people. Each ethnic group might be viewed as a branch on a genetic tree, with the branch length signifying the magnitude of the genetic difference between daughter populations that had split from a common parent. If the branches on such a tree grew primarily from genetic drift, their length would point to the time since fissioning. The branches could thus guide the geneticist to the trunk of the tree, rooted in a primeval homeland.

The project to construct a genetic tree was ambitious and visionary. It took twelve years to complete. Nearly two thousand populations were investigated from around the world. Hundreds of volunteers contributed their individually inherited traits from each population. The earliest major branching on the genetic tree stood out dramatically. It showed the exodus of fully modern man, *homo sapiens sapiens,* from his African cradle. Other milestones were equally obvious and exciting. They signified the moments long ago when a few daring southeast Asians hopped a raft of opportunity to Australia, when Middle Eastern hunters, maybe in the pursuit of big game, forded the Bosporus Strait to enter Europe and when a small band of Arctic adventurers followed a causeway across the Bering Strait to the Americas.

From one single region, the Middle East on Cavalli-Sforza's map, sprung all the colonizers who brought farming to Palestine, Mesopotamia, Persia, India, Anatolia, Europe, Ukraine, and even Egypt. The genetic distance between the newcomers and those left behind had grown through time. But in addition he had shown that the branchings of the genetic tree conformed quite precisely to those of the linguistic tree. Cavalli-Sforza had, in fact, confirmed a suspicion of Charles Darwin's that "if the tree of genetic evolution were known, it would enable scholars to predict that of linguistic evolution."

Lo n g before the science of genetics was discovered, Sir William Jones had delivered his famous discourse in Calcutta (in 1786) in which he pointed out that there must have been "some common source, which perhaps no longer exists" to account for "the strong affinity, both in the roots of verbs and in the forms of grammar," between the Sanskrit, Persian, Greek, Latin, Celtic, and German languages. "This could not possibly have been produced by accident," he had claimed. Henry Creswicke Rawlinson had depended on the observation made by Jones of a vast interchange of vocabulary and word sound to help him decipher the cuneiform glyphs on the forgotten scripts of lost languages. While reading *Hymns of the Rigveda,* Rawlinson had become convinced that stories, myths, and hymns could persist with great fidelity for thousands of years in voice alone. He believed that the Vedic hymns were one such example. There are two versions: one committed to writing very early on, and the other passed on orally with remarkable accuracy for twenty-five centuries through monotonous recitation by countless generations of Brahman monks before they were finally committed to writing, long after Sanskrit had died as a spoken tongue. It is probable that at the end the Brahman monks did not understand all that they were reciting.

Using methods rooted in Jones's observation that tongues as diverse as

Sanskrit, Greek, Celtic, and German must have descended from a common source, modern linguists have been busy pursuing languages no longer spoken and never written. Lost tongues are reconstructed in part from specific features in living languages. The restoration follows certain rules of observed language behavior.

The path of language evolution is often depicted symbolically as a language tree, whose trunk divides into several branches and the branches into more branches. The lowest part of the trunk represents the protolanguage, the branches the daughter languages, which in turn branch into other daughter languages. Daughter languages evolve when bands of people separate from a group with a common language and then, living in isolation, develop first a dialect and then a distinct language, continuing to evolve along each of the limbs, as for example English has changed from the time of Chaucer to Shakespeare and thence to its various dialects as spoken today. As needs arise, new words may be made up or borrowed. Branches may intertwine briefly with others of the same language tree or with branches of other language trees, often exchanging what are called loan words. Thus English contains many words borrowed from the French, mostly as a result of the Norman invasion. And the French in turn have recently passed laws to prevent the Anglo-Americanization of their language. Some branches and all their descendants have withered away. Others such as Greek have survived for millennia. Latin still lives, but only in the literature, ceremony, and daughter languages. One of the ancestral languages to emerge from the labors of the linguists was proto-Indo-European, which is viewed as the single mother tongue to all the descendant daughters now scattered across Eurasia and the Americas. The compelling questions are, where was this homeland and when did the speakers of the protolanguage start to break up?

The methods used by linguists to reconstruct ancient languages and language trees are in many ways similar to those of biologists in their effort to reconstruct life's evolutionary trees. Linguists reconstruct ancient languages by comparison among modern languages. In the words of Tomas Gamkrelidze and Victor Ivanov, "Linguists seek correspondence in grammar, syntax, vocabulary, and vocalization [that is, pronunciation] among known languages in order to reconstruct their immediate forebears and ultimately the original tongue. Living languages can be compared directly with one another; dead languages that have survived in written form can usually be pronounced by inference from internal linguistic evidence. Dead languages that have never been written, however, can be reconstructed only by comparing their descendants and by working backward according to the laws that govern phonological (sound) change. Phonology—the study of word sound—is all important to historical linguists because sounds are more stable over the centuries than meanings."

The rules of phonology, or pronunciation and how pronunciation may change or evolve, have been synthesized by comparative study of modern languages and their written ancestors, using for the most part those members of the Indo-European language family. Linguists have been able to deduce relatively consistent rules governing word change by comparing the vocabulary of modern languages from different branches of the same language tree and the changes in word pronunciation along a single branch of that tree. With these tools they have reached far back in time, thousands of years before the earliest written record of any of these languages, to propose a language tree for the Indo-European language that goes back to an original proto-language. Only a small fraction of the proto-vocabulary has been recovered in this way, but these words have been used to locate the homeland of these people.

For example, the proto-Indo-European had words for bear, high mountain, snow, beech tree, salmon, stream, and others that describe the country in which they lived. The homeland, so defined by the nature of its landscape and fauna, appears on maps in a wide range of locales that were seldom in the same spot but often overlap. It has been placed in northern and southeastern Europe, southern Russia, in the southern Ural Mountains, and Anatolia. One of the most dramatic scenarios is that the proto-language was first carried into Europe early in the fourth millennium B.C. by a people called Kurgans who were said to have been fierce warlike nomadic pastoralists. Their violent entry into southeast Europe snuffed out the light of the Golden Age.

The experts who attached the Kurgan warriors and their elaborate burial mounds to the homeland placed it on the Russian steppe in the vicinity of the Caucasus Mountains. Those who linked the dispersal of the proto-language with the Linearbandkeramik diaspora set it two thousand years earlier and farther west, on the open grasslands of Ukraine and Hungary. And those who envisioned the spread of language moving in tandem with the wave of agriculture saw the mother language as being older still. They put its first speakers on the Anatolian plateau where the agricultural revolution created a food surplus to power rapid population growth. It seemed for a while that there were as many homelands as there were linguists.

The only shared feature was a proximity to the Black Sea. Debate has been intense; after all, the linguists are using rules of phonological change derived from living languages, and some dead languages that have survived in written form, to extrapolate at least twenty-five hundred years beyond the first known inscription of any of the Indo-European languages.

Tomas Gamkrelidze and Victor Ivanov have questioned some of these rules. Gamkrelidze, professor of linguistics at Tbilisi State University in the Republic of Georgia, was perfectly situated to look at the whole concept of word sound afresh. Study of the phonology of the many mother and daughter languages of the surrounding Caucasus, the southern part of which was the provenance of the Kartvalian family of languages, demonstrated pronunciation changes that were contrary to the classical rules. Some sounds that had traditionally been thought to belong to the proto-language were relegated to later appearances in the family tree. Others originally considered to have appeared only after considerable evolution had taken place were now interpreted as part of the original vocalization of the proto-language. This reordering changed the order of branching of the family tree.

Gamkrelidze and his colleague Ivanov reconfigured the family tree of Indo-European languages with their new insight into the evolution of word sound. The roots of this tree reached further back into prehistory than those in previous reconstructions. When this revision eventually surfaced from behind the Iron Curtain, it raised many eyebrows. Those who correlated the diffusion of language with farming favorably reviewed the early age given to language's origins. They also liked its homeland, placed in Armenia and not far from their own setting in neighboring Anatolia. However, broadly speaking, the reception abroad was not enthusiastic. One well-known linguist in England, in reviewing all the classical ideas concerning the Indo-European origins, dismissed Gamkrelidze's work outright, stating that it "strayed light-years away from whatever consensus the general run of Indo-European studies has managed to achieve."

However, in 1995, ten years after the formulation of the Gamkrelidze language tree, the process was repeated by a team of three Americans, Donald Ringe, Ann Taylor, and Tandy Warrow, of the University of Pennsylvania. By relinquishing the task to computational mathematics, as Luigi Luca Cavalli-Sforza had done for the building of his genetic tree, and by numerically ranking all the millions of ways the genealogical chart for Indo-European languages could be configured, they hoped to avoid the bias that tends to creep in with less rigorous methods. Like its genetic counterpart, the branch lengths on their computer-generated language tree would be a measure of distance. But instead of the distance depicting periods of genetic drift, the branch lengths represented time intervals of linguistic change.

No sooner had the numbers been crunched than they vindicated one of the basic conclusions of Gamkrelidze and Ivanov—that the Anatolian languages were among the first to branch off the proto-Indo-European. The American team was caught off guard to see Anatolian so old. In the words of team leader Donald Ringe, "It might be fair to say that I was biased against it [the age reversal]." Almost not believing the results himself, he ran the computer program again and again, removing some daughter languages, inserting others. In every trial run, no matter how he tweaked the input, the same general picture emerged. "So you can imagine how startled I was when the algorithm kept turning up Anatolian as one first-order branch of the family, and everything else as the other first-order branch," Ringle remarked.

What Gamkrelidze had deduced and what the computer had confirmed was an initial bifurcation into two trunks—one for southerners, the speakers of Anatolian, and the other for northerners, who then divided into groups migrating into the interior of Europe, the speakers of Celtic; groups traveling to the Mediterranean coast of Europe, the speakers of Italic; and groups adventuring eastward into the heart of central Asia, speaking the ancestral form of Tocharian. Gamkrelidze placed the divide in early fifth millennium B.C., about six hundred years after the Black Sea flood.

Gamkrelidze and Ivanov could not have been aware that, because of the climate crises brought about by the Younger Dryas and the second mini Ice Age starting 6200 B.C., the Black Sea had become a giant freshwater lake bordered by the Caucasus and the Anatolian Plateau. As an oasis in the midst of an arid landscape, it apparently attracted people of diverse cultures and language families to flourish along its fertile shores, exchanging goods and ideas and bits and pieces of their languages. Words borrowed by the Indo-Europeans from other languages such as Semitic, Kartvelian, Sumerian, and even Egyptian attest to the proximity of these people. The famous Russian plant geneticist Nicolae Vavilov found impressive examples of Indo-European contributions to those other languages such as *vinograd* in Russian, *vino* in Italic, *wein* in Germanic, *wino* in Kartvelian, *wijana* in Anatolian, *wajnu* in proto-Semitic, and *woi-no* in the parent Indo-European. The residue of retsina wine found in the bottom of a pottery jar from sixth millennium B.C. villages near Lake Urmia on the Iranian Plateau indicates that experienced vintners were thoroughly enjoying the grape at that time.

As an oasis the Black Sea rim might have acted as a mixing pot, both for genes and language. Perhaps this is why Gamkrelidze was able to recognize so many words shared by the proto-Europeans, the proto-Kartvelians, the proto-Semites, and the proto-Ubaidians who would one day parent the Sumerians. It may also explain the transfer of a propensity for type B blood from settlers in southern Russia to those who would later immigrate to Mesopotamia and Egypt. The refuge for already-practicing farmers might also have provided a place to share tools, practical knowledge, seed, and livestock. As noted by many linguists, the borrowed vocabulary is especially rich in agrarian jargon. With its fertile river valleys, potential for grazing and hunting, abundant fish, and ease of communication by boat.

The crashing through of the ocean at Bosporus, permanently drowning all the fertile oases that had brought the assembly together, scattered the inhabitants like leaves in the swirling wind. Both the language tree and the genetic tree show a great fissioning event. With hardly any warning, the inhabitants abandoned homes, fields, possessions, and food to escape with family upstream or on the high seas. Little but knowledge and skill could be rescued. Ryan and Pitman believe that the Semites and Ubaids fled southward to the Levant and Mesopotamia; the Kartvelians retreated to the Caucasus; the LBK dashed across Europe, leapfrogging from one site to the next, pushing ahead their frontier for reasons never adequately explained; the Vinča retreated upstream to the enclosed valley of the Hungarian plain. Others went to the Adriatic and the islands of the Aegean. Some refugees migrated into the heartland of Eurasia via the Don. Still others used the Volga as access to the distant steppes of the southern Ural Mountains. In due course the Indo-Europeans soon occupied an arc extending from the Adriatic, western Europe, and the Balkans across Ukraine to the Caspian Sea. From somewhere in this strip the Tocharians struck out east to settle one day in the Tarim basin at the edge of what was to become the Old Silk Route.

MARK Aurel Stein's fabulous journey to the Tarim basin of Chinese Turkestan was preceded by a lesser-known but equally daring and perilous adventure of twenty-nine-year-old Sven Anders Hedin. On January 14, 1896, he and four porters left the desert outpost of Khotan in search of the ancient city of Takla Makan, hidden beneath desert sands since antiquity. The explorer struck out into no-man's-land—a sea of monstrous dunes rising to heights of three hundred feet. For weeks his party penetrated the lifeless world until they literally stumbled upon "a dead forest of sun-bleached, wind-scoured tree stumps protruding through the sand." At the edge of the forest were structures crafted not of stone or mud-brick but of hand-hewn posts and walls of reeds attached by twine to stakes and plastered over with clay. The polished interior walls were painted with colorful murals depicting both women in flowing garments kneeling in prayer and men with black beards and mustaches that were clearly not Chinese. The pictures included nautical scenes of boats sailing on a vast inland lake. Further digging into the ruins revealed docks for the boats and wood from their keels. Hedin wrote that this lost world

> was one of the most unexpected discoveries that I made throughout the whole of my travels in Asia. . . . Who could have imagined that in the interior of the dreaded Desert of Gobi, and precisely in that part of it which in dreariness and desolation exceeds all other deserts on the face of the earth, actual cities slumbered under the sand, cities wind-driven for thousands of years, the ruined survivals of a once flourishing civilization? And yet there stood I amid the wreck and devastation of an ancient people.

The mummies found in recent years are from other settlements where the obscuring sand has been cleared away. One woman with long auburn hair and distinctive Caucasian facial features lived during the middle Bronze Age. Carbon 14 dating of the wood indicates that settlers of the lakeshore may go back to the third millennium B.C. and beyond.

PROFESSOR Orguz Erol is a classical geomorphologist now retired and living in Istanbul, Turkey. He has an expert eye for landscape. Standing at a roadside lookout, he can describe in the most fluent diction all of the subtle features of shape, slope, and roughness, which tell him whether his surroundings were sculpted by volcanoes, glaciers, rivers, or wind. In the winter of 1996 an inch-thick pile of huge color photographs, measuring twenty-four inches on each side, appeared by courier at the doorstep of his home in a seaside suburb of the Bosporus Strait. They were sent from the Istanbul Technical University and had been purchased by the British Petroleum Company in London. The photos were of an alien world, taken from a height of two hundred miles in space by the Earth Resource Technology Satellite.

Erol tried to lay them out on the floor of his study to view the whole set of pictures at once. There were too many. Placed side by side they would have covered the floor of a gymnasium. So he set about his examination one image at a time, using tracing paper and a set of ultrasharp color pencils. The photos covered the whole of the Tarim basin of Chinese Turkestan, now a site of active oil exploration. According to Chinese hydrocarbon experts, the Takla Makan desert may hold in its subsurface many times the proven reserves of the United States, including Alaska and the Gulf of Mexico.

Erol could make out every individual sand dune on the desert floor, given the extremely high resolution of the enlargements and the fancy technology of the digital cameras that had been used to take the pictures. He drew in the outlines of the dune fields. Around the edges of the depression he mapped rugged mountains, deep valleys, and giant alluvial fans. During springtime flooding, the snow- and ice-fed rivers sourced in the high plateau of Tibet had delivered vast deposits of gravel and sand to these "dejection cones." During the rest of the year, all but the largest rivers were bone dry. The few that made their way onto the desert vanished beneath the sand. Erol traced each valley indenting the basin margin, including every side gully. All the braided networks of the ephemeral streams were sketched with pencil as he mustered up years of experience walking similar landscapes in Anatolia.

To Erol's amazement the toes of the "dejection cones" displayed terraces. He could discern three. When he placed adjacent pictures next to each other, the terraces continued in perfect alignment from one sheet to the next. The terraces continued around promontories that protruded into the desert. As Erol traced the set of terraces day after day, he realized that they wrapped the desert floor like bathtub rings. Clearly they were ancient shorelines of a lake that had once been as large as the Black Sea and that had entirely filled the Tarim Basin.

Staggered by the immensity of the lake, Erol began to look for clues that might reveal when it had come into existence and when it had dried up. He noted that "the highest coastline was discontinuous and not very fresh, covered in some places by patches of dunes and sculpted by gullies in others." The middle one was, in contrast, well defined. "This," he thought, "is where the lake probably resided the longest." The lowest shoreline had been short-lived. The lake had subsequently dried up to leave in its wake a sea of sand.

Using tentative correlations to lakes in Tibet, dated by the carbon 14 method, and the reports of Sven Anders Hedin, Mark Aurel Stein, and more recent Chinese archaeologists, Erol proposed that the well-preserved middle shoreline had formed during the terminal part of the last Ice Age when the rapid melting of thick ice cover over the Himalayas and Tibet had poured immense volumes of water into the depression. The Takla Makan lake had been enormous in area and depth. Its middle shoreline stood more than three thousand feet above the lowest part of the lake floor.

Commencing in Younger Dryas time around twelve thousand years ago, the water supply from rivers nearly ceased as the ice temporarily stopped its melting and the glaciers stabilized. It seemed logical to Erol that this pause in the melting brought on the final desiccation. The lowest shoreline is the youngest. It corresponds to a relatively brief respite in the arid climate around the time that farming took its foothold in Palestine and Anatolia. The final drying up of the lake was evidently not yet complete by the time of Marco Polo's journey to Cathay because the Venetian prince had spotted a remnant lake not seen by either Hedin or Stein during their later attempts to retrace his footsteps.

The Tocharians, whose very presence in those ancient times attested to the antiquity of the Silk Road artery connecting East and West, had deserted their adopted land when Marco Polo passed close to their remains. They had either been driven away by or succumbed to drought once the lakeside garden turned into a desert and the lake bed withered to a few seasonal pools. Their language and mummified corpses announce their proto-Indo-European origin. Their fate in the Tarim basin resulting from too little water stands in remarkable contrast with the inundation that thousands of years before had driven their presumed ancestors from the margin of the Black Sea lake, triggering migrations that astonish prehistorians to this day.

Those who fled the Black Sea inundation to the river valleys of southeast Europe, the steppes of Russia and points east, to Mesopotamia, Anatolia, the Levant, and Egypt did so long before the advent of writing. One might next ask how could the tales of their origins and their adventures have survived the thousands of years that passed before the spoken word was cast into a nonperishable script? It has been suggested that some prehistory may have passed down through the ages in oral form and served as a basis for the creation and flood myths. Placing the words in rhyme may have helped in keeping their accuracy. But how does this actually work? Can a mechanism be authenticated? Is it possible that any essence of a campfire tale or a cultural myth can survive a hundred generations of repeated recitation?

CHAPTER NINETEEN

The Guslar's Song

IN a mountain valley in Serbia, Salih Ugljanin, a guslar, sang his poem to an enraptured crowd at twilight. The hushed townfolk gathered around small tables in the village's most popular kafana. They listened attentively to an epic story whose telling began more than a week before. Ugljanin recited to the accompaniment of his two-stringed unfretted guitar called the gusle. Ever since his arrival in Novi Pazar many years prior to the assassination of Archduke Franz Ferdinand and the outbreak of World War I, Salih had been singing songs that he learned as a child from a blind bard.

Life in this remote pastoral setting, near a headwater tributary of the Danube River, had not changed perceptibly during the five centuries since Sultan Mehmed II conquered Constantinople and spread the Islamic faith through eastern Europe. Men, mostly illiterate, would congregate in coffeehouses to talk politics and gossip. Yet they were there primarily to be entertained by a body of narrative poetry handed down from the distant past, generation by generation. Salih, no more able to read the printed word than his audience, did not even know that the poem he was singing had individual lines of ten syllables each. He only thought of it as a stream of concepts embedded in a fixed fabric of rhyme whose phrases may change from recitation to recitation.

Tonight he was a quarter of the way through the *Wedding of Smailagic Hebo.* The themes in his poem were grouped into patterns, expressed anew with different word combinations during each recital. In turn the patterns were woven into a story with a fabric consistent across millennia of traditions and languages, "still alive on the lips of men, ever new, yet ever the same." This month of May in 1934 the poem's story pattern could have been de-

Milman Parry starting to record epic poetry recited by the illiterate guslar Salih Ugljanin in Bosnia

scribed as a "return song" with five acts: absence, devastation, return, retribution, and wedding.

At the close of the evening's performance Salih noticed in the crowd a foreign face. It belonged to a dashing young American professor from Harvard University who had arrived that afternoon in a large beat-up truck filled with strange equipment. Milman Parry rose and approached the guslar and in fluent Bosnian dialect asked permission to return the next evening and the next day and the day after that to listen to and record Salih's song.

Parry had traveled into the interior of Yugoslavia to preserve local heroic tales, sung spontaneously in the tradition of the oral poets. One such poet was Homer, who composed the *Iliad* and the *Odyssey* centuries before written Greek literature evolved. Parry wanted to learn how it was possible for Homer, believed to have been without sight from birth and thus cut off from visual clues, to compose extemporaneously a poem of fifteen thousand lines. Parry felt an urgency to his mission. In the wake of World War I, returning soldiers had brought literacy to their remote villages. As the people began to read, their interest in the oral tradition started to wither and was likely to disappear entirely.

Basic to Parry's quest was the riddle: Had one poet composed all of the *Iliad* and the *Odyssey* in their totality? Or were these stories the assemblage of many poems by many authors? Parry's working hypothesis was that only one poet and one voice were at work and that Homer had achieved his feat exactly as Salih was accomplishing his—that is by means of oral recitation. To test his hypothesis he had to record many poems sung by many different guslars. Altogether he would cut more than three thousand phonograph disks with nearly thirteen thousand different verses.

As he listened to recital after recital, Parry realized that the composer never sang a poem the same way twice but instead a new version was created with each recital. The differences between the sessions were noticeable, but the new word selections did not change the story line at all. Parry discovered the composer used a metrical or linguistic device to facilitate his recitations, to keep the story line consistent, even when the phrases changed. Parry soon developed a theory for which he is now famous. Tragically, he did not live to learn of his recognition by other researchers. Within the year he was dead, another bright star extinguished, this time by a self-inflicted accidental gunshot that felled him en route to delivering the first products of his research to an international congress.

Parry had discovered the linguistic aid was something he called a "formula," which he defined as "a group of words regularly employed under the same metrical conditions to express a given essential idea." The purpose of the formula and the fixed trochaic pentameter rhyme was to give the composer some standard phrase he could sing while he caught his breath and regrouped his thoughts for the verses that lay ahead. Often the formula consisted of epithets strung before or after proper names, such as "Zeus the cloud-gatherer" or "long-suffering, brilliant Odysseus" from the *Odyssey*. They were not limited to gods or people but were attached as well to nouns and verbs of action. Advocates of Parry's oral-formulaic theory cite an example phrase—"when the lovely dawn shows forth with rose fingers" —taken from Homer and selected by the poet to convey a sense of time. This phrase or a version of it is repeated over and over again in the *Odyssey*, seemingly to aid the poet's process. Parry had found no less than twenty-nine formulas in the first twenty-five lines of the *Iliad*.

Although the guslar composed a fresh poem each time he sang his ballads, when questioned he would assert that his latest rendition was "the truth exactly as he heard it." Salih finished his *Song of Baghdad* with the phrases: "I say! In this way I have heard it, in this way to you I have told it."

Parry's work was continued by Albert Lord, his accomplished associate at Harvard, who undertook the Herculean task of transcribing, decoding, and cataloging all of Parry's recordings. Traveling back to Yugoslavia, Lord confirmed that the guslars had not merely learned their songs by rote. Hearing again one song that Parry had recorded at a performance seventeen years earlier, Lord found that the new and old versions were distinct, yet identical in all the hundreds of formulas used to weave the story into a unified epic fabric.

Lord found an embodiment of myth within the stories, possibly of more esoteric power than even the guslar was conscious of communicating. These myths were universal and survived because of their continuing influence on human culture. They served several functions, among them history, teaching, and entertainment. In an oral culture, encyclopedias, dictionaries, history books, and novels did not exist.

Lord recognized a symbiotic relationship between the poem and the poet. He found that each poem needed a good poet with skill and commitment who could keep the poem vital without altering its central theme. The epic poet might choose to sustain what was important to him, but he also had to satisfy his audience. Consequently, the poet needed a powerful poem, one whose survival through the ages was assured by the potency of its message. Such a poem was one that had a story line permeated by myth. Experts have argued that the myth was unavoidable. It was the most essential element to keep us listening attentively. The oral tradition was the constant repetition of these myths. The most fundamental value of oral poetry is the myth that it passes down through the generations. Even a song considered to be a retelling of times gone by has a myth at its core. The myth resuscitates the perishing history and keeps it breathing long after the legend has been declared a fantasy of the dead past.

Lord wrote several essays dealing with the issue of the myth and history. In them he argued that myths survive the changes of tradition better than the events portrayed. In other words, story patterns reign supreme over historical facts. He wrote, "The patterns must be suprahistorical in order to have such force. Their matrix is myth and not history; for when history does have an influence on stories it is, at first at least, history that is changed, not the stories." From his analysis of the Trojan war as history, Lord commented, "Fact is present in the epic, but relative chronology in the catalogue is confused. Time is telescoped. The past of various times is all assembled into the present performance. Oral epic presents a composite picture of the past." In one example of the exquisite "fact" that can survive through the ages as decoration to a story, Homer describes the weapons of terror used by Odysseus. They consist of a two-edge sword and "a helmet fashioned of leather; on the inside the cap was cross-strung firmly with thongs of leather, and on the outer side the white teeth of a tusk-shining boar were close sewn one after another with craftsmanship and skill; and a felt was set in the center." This same boar's tusk helmet is portrayed in a Minoan Age (circa 1630 B.C.) fresco on the wall of a residential house in Akrotiri on the Aegean island of Thera, buried beneath the ash and pumice of an enormous volcanic eruption six centuries before the alleged Trojan war.

Lord taught a course at Harvard in oral epic tradition. He included not only an analysis of the living Yugoslav epics but the ancient ones of the Anglo-Saxons, Scandinavians, Russians, Greeks, and Persians, including the Sumerian and Akkadian creation epics and the epic of *Gilgamesh*. The work was not presented in chronological order but proceeded from what was thought to be the more primitive to the more sophisticated forms. Lord considered the Gilgamesh epic especially sophisticated because its central theme was the human experience and not that of the gods.

J O H N Miles Foley presently directs the Center for Studies in Oral Tradition at the University of Missouri. He coordinates and edits the research of thousands of modern scholars investigating traditional literature around the world, ranging from the songs of African "giots" today to the poems of Sumerian priests in the Mesopotamian past. In every case Parry's hypothesis has borne fruit. It has uncovered the presence of formulae in poems, songs, recantations, speeches, wedding vows, prayers, hymns, and more.

The "return song" story pattern that Parry listened to in the coffeehouse of Novi Pazar has since been found in a large number of traditions spanning three thousand years of oral passage. Does this discovery mean that many languages, cultures, and civilizations branched off from the trunk of a single culture, or prehistory, and carried forth the story patterns and critical themes of the "primal civilization"? Or, instead, are the "primal" story patterns consistent across cultures and traditions because they address issues central to humankind everywhere such as love, death, birth, and life—issues that every society must understand and explain for its own cultural survival? These two questions lie at the center of debate in the contemporary study of mythology, particularly to the extent that the myths have spawned a written literature subject to critical analysis and deconstruction.

The treasure from the royal library at Nineveh contains some of the best preserved myths and legends transmitted from the deep recesses of human memory. Some tablets extend back to the dawn of writing five thousand years ago, little modified by subsequent scribes. The collections we have today are far from complete, having suffered greatly from inept digging and from the mandates and tastes of the original archivists and their librarian kings.

Henrietta McCall, a specialist in Mesopotamian literature, has asked, "How representative of the library tradition as a whole is the literature that has survived?" As a partial answer she and colleagues at the Ashmolean and British museums have pointed out that the libraries themselves contain the clues. The administrative records are remarkably well organized and indicate that new acquisitions were sought with vigor. Works were organized according to title and genre. The records show that private collections in Babylonia were extensively incorporated into the Nineveh public library, especially after the sacking of Babylon in 648 B.C. In fact, King Assurbanipal himself supervised aspects of the acquisitions.

In some Assyrian repositories the edges or last column of the tablets were labeled with identifying marks that we can liken to an index. They include a title, the names of the owner and the scribe, the date in the context of a king's reign, and other comments. They sometimes included a curse for anyone defacing or failing to return the tablet. The very existence of Mesopotamian literature presumes that there were individual authors to the works. However, their names are, for the most part, never mentioned. In Assurbanipal's library a compendium of authors for some texts lists gods, celebrated heroes, and those of great antiquity—all the while asserting a legacy from prehistory for the great legends. The vast majority of different versions of literary works that have survived are compilations from many sources. Furthermore, many of the sources used were also derivative in nature. McCall describes Gilgamesh, in which George Smith discovered the account of the Chaldean deluge, as a "stitch-up job . . . with joins that are less than smooth."

In looking for authorship of our earliest preserved literature, as was the case in Greece for the *Iliad* and the *Odyssey,* the question arises as to whether or not these works are descendants of oral tradition. The creation story that George Smith was deciphering at his untimely death specifically states that it was to be recited. One Sumerian poem, *Dumuzi's Dream,* has received a careful scrutiny of its presumed oral heritage in light of the Parry-Lord theory. Dumuzi was a Sumerian cupid, bed partner of Ishtar, the goddess of love, sex appeal, and war. He is immortalized in many different stories and texts, often with identical lines and the same motifs. The premise for considering that the Sumerian poems were oral in origin stems from the observation that they used a common set of formulas, despite their existence in widely differing recensions. Within a single version they generally employ the same traditional phrases and fixed sequences of words to express the same idea.

Repetition is integral to an erotic Sumerian love song to a king entitled *The Honey-man,* which is overflowing in Freudian imagery and metaphor. *Lady Chatterley's Lover* pales beside it.

He has sprouted, he has burgeoned, he is lettuce planted by the water,
My well-stocked garden of the plain, my favored of the womb,
My grain luxuriant in its furrow, he is lettuce planted by the water,
My apple tree which bears fruit up to its top, he is lettuce planted by the water.

The honey-man, the honey-man sweetens me ever,
My lord, the honey man of the gods, my favored of the womb,
Whose hand is honey, whose foot is honey, sweetens me ever.
Whose limbs are honey sweet, sweetens me ever.

My sweetener of the navel, my favored of the womb,
My sweetener of the fair thighs, he is lettuce planted by the water.

Another reason for believing that Sumerian poems were oral in origin is that they possess instances of narrative inconsistency. Details of descriptive settings and the chronology of events are muddled, especially in the longer tales such as *Gilgamesh* that were compartmentalized into discrete episodes. These vagaries may reflect a guslar's choice whereby traditional story episodes may be combined out of order in the poem as the poet recomposes it.

Gilgamesh is so embroidered with legendary themes and mythical motifs that its historical foundation requires elaboration. According to the Sumerian king-list, Gilgamesh was the fifth king of the First Dynasty of Uruk (biblical Erech), a city-state on the banks of the Euphrates River during the Second Early Dynastic Period of Sumer (circa 27th century B.C.). Within two hundred years, Gilgamesh was widely revered as a god in Sumer. The adaptation of his recorded exploits from the oral recital to cuneiform inscriptions may have been initiated about this time. But even then continued oral performance by a class of illiterate storytellers was almost certain since the complex system of cuneiform signs kept writing within a tiny elite of professional scribes. A thousand years later the epic was dispersed abroad to Anatolia in the Hittite language and to the Canaanites of Palestine in Hurrian prose. Although it is impossible to know the degree of modification from inherited tradition, ritual details are strikingly similar in content and style among all the versions. Apparently the guslars, in the presence of a powerful poem, had been diligent in the faithful perpetuation of their ancestral myth.

As Milman Parry discovered, for a myth to survive unscathed from repeated recitation, it needs a powerful story. A narrative of human history from its origins to its present is, on its own, compelling, especially when interleaved with great events that might be interpreted as the random whims of the supernatural gods (chaos) or deserved punishment (determinism) for one's own transgressions, and which lead to heroic struggle for survival and eventual understanding of one's place within the wider cosmos. Oral tradition tells such stories. But so does the decipherment by the natural scientist who works from a text recorded in layers of mud, sand, and gravel from the bottom of lakes and seas using all the tools and principles of physics, chemistry, and biology. The scientific plot can then be given richer detail and new themes from the supporting contributions of the archaeologist, the linguist, and the geneticist.

IV

The Flood Stories Told

lies the town of Anadoluhisari and across Rumelihisari, the remains of two
forts built by Sultan Mehmet II during the siege of Constantinople. Darius is
reputed to have bridged the Bosporus at this point.

Ships ease by heading north and south, and at dawn the first of the
ferryboats appear. In a few hours they will grow to a swarm plying back and
forth and up and down. This gentle awakening to a new day belies the
dramatic events that occurred there over seventy-six hundred years ago,
drastically altering the landscape of the Bosporus and the entire area of the
Black Sea and forcing a diaspora that changed the course of human history.

But the story begins long before, back at the beginning of the last glacial
cycle, 120,000 years ago when the Earth's climate and the level of the seas
was about the same as today. From that point in time and for the next
100,000 years waters evaporated from the oceans and, transported by the
winds, fell as snow on the near Arctic regions, gradually accreting and
compressing into sheets of ice that were in some places up to two miles
thick. Twenty thousand years ago at the zenith of this accumulation so much
water had been withdrawn from the oceans that sea level was four hundred
feet lower than today. Massive glaciers covered the entire northern half of
North America, all of Scandinavia and northern Europe, and the northern
edge of Eurasia. All the high mountains of Europe, Asia, North America, and
South America were sheeted with ice down to their lowermost valleys.

Modern man was there in Europe and Asia to witness and survive the
extremity of this glacial episode. Having emerged from Africa about one
hundred thousand years ago, the Moderns spread across Asia, rafted over to
Australia, and finally entered Europe about thirty-five thousand years ago.
Hunter-gatherers, they made stone tools and lived in temporary shelters and
caves much like their very distant cousins the Neanderthals. But they
brought with them qualities that their predecessors lacked, in particular an
extraordinary ability to adapt and innovate. Their inventiveness is reflected
in the rapid evolution of their hunting and survival skills.

Twenty thousand years ago the great glacial meltdown began. Torrents
of frigid waters raced to the sea, which slowly began to rise. Gradually the
huge icy burden was removed from the land. In northern Russia rivers
choked with meltwater flowed southward across the steppes and eventually
spilled into the Black Sea's Ice Age lake. The icewater filled this lake to a
level where it entered the Sakarya River, formed an estuary and advanced
into the interior of Anatolia. Fifty miles in from the coast, this narrow and
winding arm of the expanding lake found an outlet to the Sea of Marmara,
having intercepted a cleft in the bedrock wrenched open through the grinding action of the North Anatolian fault (the locus of numerous large-

magnitude deadly earthquakes delineating a zone of shear along which two of the earth's plates still slide past each other at the rate of inches per year). Exploiting the crushed and permeable rock in this crack, the meltwater passed through to the Mediterranean Sea. In the process, the Ice Age lake freshened and became potable for humans and animals.

Those Moderns who lived in the north followed the great herds of mammoth, reindeer, and other large herbivores as they migrated across the tundra in front of the retreating glaciers. They lived in hunting camps for several months a year. On the Russian steppes where wood was scarce, they framed their lodges with mammoth bones, interlocked, tied together, and covered with skins. Meat was stored in pits dug into the frozen ground. They sewed skins into pants and boots, and made jackets with hoods.

In the more temperate areas of Europe and Asia, the Moderns lived in round and oval huts of reeds or poles and skins sometimes built over a shallow pit. Fish were caught with hook and line and in traps and nets. They invented the bow and arrow, and the throwing stick that doubled the range of their hunting spears. Some may even have smoked and salted meat and fish. But life remained a day-to-day struggle for survival.

As they met the challenge of this most difficult and demanding environment, there emerged from their creative impulses a most remarkable proliferation of art objects, often rendered with astonishing skill and finesse. Frequently quite decorative, many of these objects may have been created for aesthetic reasons only. But most seem to have been made for mystic and cultic purposes. Beads and amulets were carved of ivory, stone, and shells, and used to decorate clothes and to make necklaces and bracelets. Marvelous statuettes were sculpted of ivory, bone, and stone. Throwing sticks were carved into stylized images of horses, deer, and other animals they preyed upon. Weapons for the hunt, tools, and decorative objects such as bracelets and necklaces are found in graves with the body of the deceased as if these would be needed in some afterlife.

The most spectacular of this "primitive" work are the cave paintings found for the most part in France and Spain. Usually rendered on the walls of chambers deep underground, they have been painted and scratched into the rock with extraordinary grace and skill. There are striking monochrome and multicolored paintings and line drawings of buffalo, deer, mammoths, horses, lions, hyena, and other animals and birds. Painted dots, chevrons, and silhouettes of hands are often interspersed among the figures of the animals. There are a few human portraits and several graceful drawings of the female figure. In many cases pictures have been superimposed on one another as if the act of drawing was important rather than the drawing itself.

This artwork, sometimes quite realistic and at other times very abstract, has been found on the walls of caves and on cliffs, and even in a cave along the shore of the Mediterranean, long submerged by the rising sea level, and everywhere the Moderns roamed.

A very clear and persistent manifestation of the spiritual and mystical nature of the Moderns are statuettes of women, interpreted almost universally as a symbol of fertility cult, found at sites throughout Europe, the Middle East, North Africa, and as far east as Lake Baikal in Siberia.

FIFTEEN thousand years ago the glaciers were in full retreat, pouring millions of gallons of frigid meltwater into the rivers across North America and Eurasia. The colossal weight of these ice sheets impressed itself on the Earth's surface like a heavy object placed on a soft mattress. The weight on the mattress does not fill the dent entirely but is surrounded by a moat. Likewise the glaciers are often bordered by moats that trap meltwater, debris, and large chunks of ice that have broken away. Because in the northern hemisphere it was always warmer to the south, the glaciers melted mostly along their southern edge. As they retreated, the moat followed northward, across the steppes, trapping water and diverting flow to follow the channel parallel to the glacier front. By 13,000 B.C. the ice had withdrawn so far north that the flow of meltwater to the Black Sea had almost ceased.

Europe was gripped by a return to the rigors of the glacial climate 12,500 years ago, an event known as the Younger Dryas, which lasted for a thousand years. Temperatures fell and the rains became scarce throughout southwest Asia, Europe, and Africa. Glaciers advanced in the high mountains. Lakes in Africa and Anatolia dried up. Precipitation in and around the Black Sea was low, so inflow was reduced to the point that the loss of water by evaporation from its surface exceeded the water received from the rivers and rainfall. The water level began to drop until it had fallen below the Sakarya outlet. Outflow ceased, and the Black Sea became an isolated lake.

The Sakarya channel, no longer connected to either the lake or the sea, slowly collected mud and debris brought in by torrential seasonal rains and the flooding of its several streams. This debris built up slowly, forming a natural earthen dam. As the lake slowly drew down, it exposed an old shelf, a thick accumulation of the remains of marine organisms and rich sediments brought by the many rivers. The retreating waters, driven by winds and tides, sloshed back and forth, removing the silt from the uppermost layers of the sediments, leaving only the fragile shells of the delta mollusks, broken and bleaching in the sun. Sunbaked cracks filled with sand and seed of the wild wheats native to the area, some of which took root—especially in the more moist depressions and valleys. New sinuous valleys were cut right out to the shoreline, and detritus carried by the rivers in these valleys was deposited at the newly lowered lake edge, building new deltas that were bordered by the natural levees created during the occasional overflow. These valleys and deltas—with their rich soils, nurtured by the constant, if sluggish, flow of water and with abundant fish life in the rivers and at the edge of the sea—became an ideal refuge for man and beast.

In the Near East many bands of hunter-gatherers had adopted a more sedentary way of life, constructing permanent villages, hunting and fishing locally, and gathering fruits, nuts, and wild wheat and barley, which they later learned to cultivate. With the coming of the Younger Dryas, however, and the sudden change to a cooler and arid climate, these resources disappeared. Jericho was deserted, as were many other villages. The plains of Ukraine and southern Russia reverted to steppe desert. Tribes crowded near oases where game and water were plentiful, such as at the rim of the Black Sea lake. There on the deltas and the river terraces, at the edges of the lagoons—perhaps due to the accidental scattering of some of the wild seeds they had harvested—they learned the lessons of sowing grains, the first step in farming. They also traded food, goods, and ideas with others around the lake.

The Younger Dryas ended 11,400 years ago, as abruptly as it had begun. Warmth and rain softened the harshness of the surrounding countryside, and over a period of a few hundred years the landscape was revitalized as game and the wild fruits, nuts, and grasses returned. People began to move away from the oases, taking with them the newly acquired skill of farming. They spread into Anatolia, the Levant, and northern Mesopotamia, flourishing in the valleys that were well watered again and along the shores of lakes.

In 6200 b.c. this tranquil existence was once again disturbed by another mini Ice Age that seized the Northern Hemisphere. Temperatures dropped and rains were meager. A wave of aridity again swept across southeast Europe, Ukraine, and southern Russia. The lakes and rivers of Anatolia, southwest Asia, and southeast Europe shrank. Many farming villages in Anatolia and along the Fertile Crescent were abandoned, while others dwindled. Communities of people, many of whom were now farmers, retreated to the watery patches, to the few rivers that still flowed and to the rim of the Black Sea.

Sea level was still below the level of the divide that separated the Bosporus valley from Marmara. The Black Sea remained an isolated lake. Most of the people who came to the lake edge this time were farmers who cultivated the river valleys and deltas. Here again they traded around the lake edge, now using small boats, speaking different languages—Proto-Semitic, Proto-Indo-European, Proto-Kartvalian, and others—exchanging goods, obsidian, leather, pottery, herbs and essences, and borrowing words for new things and ideas. Some on the edge of being hunter-gatherers and herders adopted a new way of life and learned to farm from their agrarian neighbors. A Sumerian myth declares that they were given their knowledge and culture by "seven wise men who came from the sea." Eventually on one of the deltas someone may have diverted the water through a natural levee and invented irrigation.

Relief came around 5800 B.C. as the rains and warmth returned and some of the lakeside dwellers, such as people called Halaf, left the basin and reoccupied a few of the abandoned sites to the south. By 5600 B.C. the ocean had risen to a height where it stood poised to invade the Bosporus valley, and plunge to the Black Sea lake five hundred feet below. Driven by the wind and tide, the waters must have repeatedly washed up onto the top of the divide to fall back, leaving damp patches on the soil, until a final surge began to flow continuously across and down the slope toward the lake, finding old gullies and dried streambeds in the rough ground between the trees and around the litter of boulders.

Reaching the ancient shelf below, the water meandered across its flat surface, trickling into old channels long dry, formed small lagoons, and gradually cut its own course, at last flowing over the edge and down the gentle slope to the lake below. The rivulet became a gentle brook, flowing ever more swiftly, scouring and tugging more forcefully at the bottom and walls of its channel. Within days its gentle murmur would have grown to a roar as the stream became a wildly turbulent river, cutting into its banks, pulling trees and large chunks of earth into the maelstrom.

The soil and debris that had once dammed the valley were quickly swept away, and the water, now several tens of feet deep, was a thundering flume twisting and churning with rubble as it clawed at the soft rock walls that now and then collapsed. The debris-laden water ground into the bottom like a rasp, cutting deeply into the bedrock itself. The deeper it cut, the faster it flowed, and the faster it flowed, the faster it cut until it had gouged a flume at least 280 feet and up to 475 feet deep. Ten cubic miles of water poured through each day, two hundred times what flows over Niagara Falls, enough to cover Manhattan Island each day to a depth of over half a mile.

Most if not all the fish life in the lake died in the strange salty water. New life that flushed in from the Mediterranean at first perished in the maelstrom, but after some days, when the torrent had abated, it made it through and quickly took over.

The level of the lake began to rise six inches a day, immediately inundating the deltas and invading the flat river valleys—moving upstream at as much as a mile each day, without pause hour after hour, day after day, drowning the less agile, forcing all else upriver or up onto the desertlike plateau through which the valley had been cut.

It is hard to imagine the terror of those farmers, forced from their fields by an event they could not understand, a force of such incredible violence that it was as if the collected fury of all the gods was being hurled at them. They fled with family, the old and the young, carrying what they could, along with fragments of the other languages, new ideas, and new technologies gathered from around the lake.

Farmers, called Vinča, makers of lovely wattle-and-daub houses and fine incised pottery, appeared abruptly on the plains of Bulgaria and up along the valley of the Danube. Other refugees crossed from the Black Sea to the Aegean to settle on some of the islands such as Samothrace, crossing over as far as the Dalmatian coast. Some Linearbandkeramik fled up the Dniester River and then rapidly to the west across northern Europe as far as the Paris basin, displacing peaceably or by force the indigent hunter-gatherers. They brought with them their longhouse building style, their ceramic pottery decorated with linear bands of incised grooves, and their agrarian ways. They may have been Indo-European speakers, but others who certainly were moved out to the north, up through the river valleys of the Dnieper, the Don, and the Volga, spreading in an arc from southeast Europe to the Caspian Sea and beyond. It was around the northern Caspian Sea that they first domesticated the horse, on the backs of which they stormed into eastern Europe sixteen hundred years later.

The speakers of the Semitic tongues climbed through the hills to the south, up creeks and streams and over the Anatolian plateau, scattered wide by the complex of deep valleys and mountains. In Anatolia some of these desperate peoples laid siege to a few small villages, which were then burned to the ground. Many deserted farm villages of the Levant were suddenly reoccupied, and strangers with an advanced farming technology and domesticates of foreign origin settled in Egypt on the Nile Delta. Many more of the Halaf appeared along the northern fringes of Mesopotamia and ventured southward into its arid valleys farther than any farmers had dared before.

Some of the Semitic peoples who crossed southward through eastern Anatolia and speakers of Caucasian languages who fled from the eastern end of the Black Sea to the south or up the Rioni Valley and around to the south, drifted down along the eastern side of Mesopotamia, and settled in the foothills of the Zagros Mountains. They, too, were farmers.

A few of these, called Ubaid, speakers of a tongue later to be known as Sumerian, ventured to the middle of the southern Mesopotamian alluvium, a region where the rainfall was only four inches a year and where the only natural resources were the extraordinarily rich soil and the Tigris and Euphrates rivers that bound the plain. These people, who knew how to irrigate and may even have used a light plow, flourished. Irrigation here required canals and, hence, social organization was needed to design, plan, and maintain the canals.

The exceptional fertility, the presence of a limitless supply of water for irrigation, and the growing web of canals for transport meant that the few could provide for the many. Prosperity led to more prosperity, and soon one of the world's great civilizations emerged as a people and culture known as Sumerian succeeded from their ancestors, the Ubaids. There flowered among them an amazing pantheon of gods, one for every need. A very superstitious and fatalistic people, they believed in predestination but also that the future could be revealed by divination. They sought a cause for all events among the gods. Small wonder that their entire history was recorded in myth. By 3000 B.C. these people had invented writing. Using wedge-shaped styli to inscribe symbols on tablets of wet clay, they recorded the everyday events of their lives and immortalized their myths, religious beliefs, and practices.

The flood continued long after the human population had fled. With awesome unabated violence the waters poured through the Bosporus day after day. The rising waters filled the river valleys and dry channels of many old streams and brooks of the old shelf to the north, forming a glittering web of canals. For twelve months the tumultuous rush of water continued undiminished until the level of the lake had risen 180 feet, to the lower surface of the flume. As it continued to rise, the rate of flow slowly began to diminish. Still, during the next twelve months it would rise another hundred feet. It crested the old shelf edge and began its race toward the present shoreline, pushing all life before it. Rolling over the scrub bushes, desert grasses, and small trees, it inundated the sparse vegetation, sand, dirt, and fragments of shells left behind eons before to bleach in the sun. Everywhere the encroachment of the floodwaters was so rapid that whole regions that had been dry were covered by ten or more feet of water within days. There was no continual wash of waves to form sandy beaches, but in this suddenly deep and quiet water, the sediments, shells, and skeletal structures of very small sea creatures and particles suspended in the water floated down like dust, covering everything with a uniform layer without regard to the underlying topography.

With these foreign waters came migrant species from the Mediterranean, now able to colonize in the newly saline water and on the shelly bottom. On one of these newly flooded muddy shallows the immigrant cockle, *Cardium edule,* having survived the rush at the Bosporus, proliferated, lived, and died. The delicate shells of these first colonizers settled on the bleached shelly surface of the old shelf, to be picked by tweezers from sediment cores 7600 years later.

All around the lake the tentacles of salty water reached up the rivers and creeks, pushing farther and farther inland each day. Along the south edge of the lake the waters quickly swamped the deltas and followed the valleys into the fringing mountains of Anatolia, chasing life up into the hills.

After two years, when the lake level had risen 330 feet, the waters entered the Kerch strait and shortly thereafter reached the Azov plain, which had been abandoned long before by humans. It would be several more years before the basin was completely filled, creating the Sea of Azov, so that its surface, like that of the Black Sea, was at the same level as the Aegean and Mediterranean seas, beyond. Sometime afterward the flow through the Bosporus slowly changed to its present state with fresher, lighter Black Sea water flowing out at the surface and the heavier Mediterranean water flowing in along the bottom. Once this change had occurred it would be impossible for surface-dwelling creatures to travel northward into the Black Sea except as stowaways in the bilgewater of ships. And most vessels could not have made this journey until the discovery of the northward-flowing bottom current.

To those who fled to Mesopotamia, in particular, the mythic and oral tradition of the event would have been reinforced by the frequent if irregular floods that occur there. The myth lived on, sung and chanted at ceremonies and around caravan campfires by generations of guslars and storytellers. Each year that the floods came was probably the occasion to retell the story of that ancient time when the Great Flood destroyed all people but one single family from whom all were descended. Elaborated and modified to conform to the more familiar geography of Mesopotamia, it still retained its basic theme: a warning, a violent flooding, the escape of a family, the apparent inundation of the entire world, the apparent retreat of these waters, and the landing and salvation of these people.

CHAPTER TWENTY-ONE

Other Myths

SIGNS of unsettled people on the move appear in the archaeological debris of southwest Asia and Europe immediately after the Great Flood. Linearbandkeramic people and other refugees fled up the river valleys of Europe and western Asia. Desertions, occupations, and conflagrations occurred in Anatolia, migrants descended into the fringing hills of the Fertile Crescent, new farmers appeared on the Nile Delta and along the tributaries of the Kura River, and peoples called Ubaid brought irrigation with them and cultivated southern Mesopotamia for the first time.

Yet the linkage of this apparent diaspora with a permanent flooding of the Black Sea is only a working hypothesis. Short of finding the remains of Neolithic settlements beneath the mud of the present Black Sea shelf, no archaeological observation can prove a human occupation of the now-submerged landscape.

Perhaps in the future the marine archaeologist will locate tells, the drowned remains of towns with their water-soaked mud-brick walls, fire-baked hearths, postholes, and stone tools up along the drowned river valleys and the ancient lake's rim. And perhaps even recover artifacts linking those people to refugees who fled. All Ryan and Pitman could now assert was that settlement around the shore and in rivermouth deltas was possible and even probable, given the well-developed farming culture in the vicinity of the largest body of fresh water in the region during a period of sustained aridity. The known Neolithic farming sites were invariably settled on water-retentive alluvial soils of the same type cored by the Soviets all across the now-submerged broad continental shelf of Bulgaria, Romania, Ukraine, and Russia.

With not only a real Black Sea flood to work with, but also corroborating evidence of concurrent human migrations into Europe, Palestine, and the alluvium of Mesopotamia, Ryan and Pitman had to address seriously the question John Dewey had impulsively posed twenty years earlier: Could this violent rush of salt water into a depressed freshwater lake in a single catastrophe have been the inspiration for the flood mythology?

When the Reverend William Buckland assumed his role as the leading authority in the natural sciences at Oxford in the early nineteenth century, his Anglican faith served as the basic foundation of truth on which all else was derived. Noah's Flood had occurred; God said so. Therefore Buckland perceived his own task and purpose to be the search in the natural world for the effects of the global submersion. The loose sedimentary deposits overlying much of the regional bedrock of his native Scotland were in his eyes either "diluvium," laid down by the flood, or alluvium that he could see for himself to be the contemporary action of streams in high flood stage. When human artifacts or bones appeared among the skeletons of coeval and now-extinct Ice Age mammoths or cave bears in the so-called ossiferous caverns of the British Isles, it was proof that man had nearly perished at the hands of God in the Genesis account. Buckland was no less honest or rational than modern-day scientists. He simply launched his intellectual inquiry from a different starting point. But when he was ultimately confronted by the enormous contradiction between the glacial origin of the "diluvium" and the flood origin, he and Charles Lyell converted to the more plausible Ice Age interpretation.

Whereas Buckland started with the Biblical flood which he assumed to be true, the method here is to start with the catastrophic pouring of salt water through the Bosporus gorge. The rapid and permanent filling of the Black Sea is taken as a fact, deduced from the scientific evidence. The strategy is to search the mythology to check for any credible Black Sea fingerprints. Are there clues within the myths that might point to a Black Sea origin?

George Smith believed the stories and myths from King Assurbanipal's royal library to be genuine traditions, some tales of pure romance but others compiled to account for natural phenomena. He wrote, "The details given in the inscriptions describing the Flood leave no doubt that both the Bible and the Babylonian story describe the same event, and the Flood becomes the starting point for the modern world in both histories." But what literal credence can be given to these "details" in the inscriptions? Even if the first storytelling about the catastrophe was true to the specific location and circumstances of the event, would not most of the descriptors have been lost or discarded over the ages as the content of the legend was diluted, distorted, or otherwise embellished in the songs of a hundred or more generations of guslars? Not until the tales were written in cuneiform script on the clay tablets would the substance of the legend be stabilized and immortalized for posterity.

Today many scholars argue that myth has little connection with the past. Myth has been perceived as a dream image, "a little hidden door in the innermost and most secret recesses of the soul." The late Joseph Campbell reasoned that "the material of myth is the material of life, the material of our body, the material of our environment, and a living, vital mythology deals with these in terms that are appropriate to the nature of knowledge of the time." This in no way argues against the conclusion that some myths may have a firm foundation in historical events, particularly one as dramatic and devastating as the flood.

George Smith had struggled with the tablets that described the contest between Gilgamesh, Enkidu, and the monster Humbaba, guardian of the forest. The partly assembled cuneiform text described Humbaba as living "in the midst of a region of *erini* trees, where there were also trees of the species called *Survan*." Smith translated these as "pine" and "cedar." He noted that "in one inscription Lebanon is said to be the country of *survan*, in allusion to its cedar trees." All subsequent translators have named the same species, but the location of the Humbaba's dwelling has been a subject of some discussion. The earlier Sumerian versions allow the possibility that it might have been to the east of Mesopotamia, near the Zagros Mountains, due to a reference to the sun god, Shamash, which is thought to mean sunrise. However, in the later Akkadian texts, the Cedar Mountain is explicitly located in the northwest, in or near Lebanon, and the reference is to sunset. Humbaba's Cedar Mountain is inferred to be distant from Uruk. Gilgamesh and Enkidu began their trek on the banks of the Euphrates and crossed seven mountains before they came to the gate to the forest. "The hugeness of the cedar rose in front of the mountain, its shade was beautiful, full of comfort. . . . There Gilgamesh dug a well before the setting sun."

Since the slaying of Humbaba had been given a physical setting by Assyriologists, Ryan and Pitman concluded that it was legitimate to consider other, perhaps more vague geographic references in Gilgamesh's journey to try to find the ancient Utnapishtim, the survivor of the flood, the Akkadian Noah, from whom Gilgamesh wishes to learn the secret of eternal life. After roaming the hilly steppe Gilgamesh arrives at a mountain range whose peaks reach high into the sky. Scorpion people keep watch at its gate. Gilgamesh is told that "no one had crossed through the mountains, for twelve leagues it is darkness throughout—dense is the darkness, and light there is none."

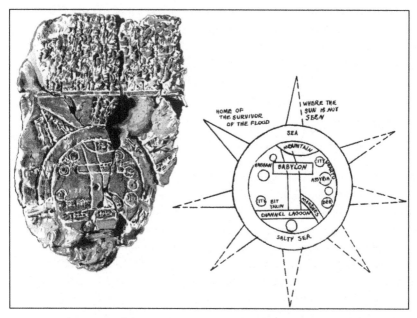

A Sumerian map of their world, depicting Babylon, the Tigris and Euphrates,
the salt marshes, the Persian Gulf, the land without sun, and the alleged
home of Utnapishtim, survivor of the flood

For twelve leagues he has to walk and climb; it is impossible to see anything
either ahead or behind; the weight of the blackness is pressing in on him, as
if following the way of the sun's night journey. Gilgamesh treks literally
through the depths of an "underworld."

A geographer imagining this passage in the *Gilgamesh* epic would posit
a route northward from Mesopotamia into the headwaters of the Euphrates
River, passing first through the hilly grasslands of the Fertile Crescent and
then up into the perilous mountain peaks surrounding the region of Lake
Van in Anatolia, once the wild kingdom of Urartu. These are the Taurus
Mountains, whose snow-capped peaks reach elevations of fourteen thousand feet, a formidable obstacle. In remote mountain villages the Kurds still
weave patterns of scorpions and fantastic beasts into their wool rugs. The
Urartu were a confederation of indigenous tribes who tangled fatally with
Sargon II, king of Babylon, in the eighth century B.C. The word *Ararat*, the
name of the mountain on which Noah's ark settled, comes from the Urartu;
Utnapishtim is the son of Ubartutu, a possible reference to this geographic
region.

The ancient capital city of Çavustepe was carved out of volcanic bedrock high on a promontory above the pure turquoise lake. Its intricately fitted stone walls of black basalt still show the traces of an Assyrian attack by King Sargon. At that time the plains below and the valleys all around were covered with forest. Sargon described "high mountains covered with all kinds of trees, whose surface was a jungle, whose passes were frightful, over whose areas shadows stretch as in a cedar forest, the traveler of whose path never sees the light of the sun."

When three hundred years later Xenophon led his mercenary Greek army of ten thousand northward from Babylon across Anatolia to the shore of the Black Sea in order to escape slaughter by pursuing Persians and the local natives, he, like Sargon II, had to lead his soldiers in single file to penetrate the primeval forest.

Gilgamesh finally emerged into a brilliant garden. "Before him there were trees of precious stones, and he went straight to look at them. The tree bears carnelian as its fruit, laden with cluster (of jewels), dazzling to behold." Is this the glitter of the sun on ice high in the mountains?

Finally he comes to the sea, the Black Sea according to Ryan and Pitman, and to the abode of the "alewife," a tavern keeper; Gilgamesh asks how to cross the waters to see Utnapishtim. The sea is inhospitable, he is told. "Gilgamesh, there has never been a crossing. . . . Everywhere the waters of death stream across its face." Not from the beginning of time has anyone been able to go across this sea of death. "Far out in the waters, forbidding the way, there slide the other waters, the waters of death."

The Black Sea is indeed a sea of death. In its surface waters there is sufficient oxygen to support a varied and plentiful fish life, but since the flood, its depths, devoid of oxygen, have been enriched in poisonous hydrogen sulfide. A lungful of this gas will kill a human. Today the interface between the life-supporting surface layer and the life-denying anoxic layer lies 450 feet below the surface. Its level is not constant. In the last few decades it has risen by more than eighty feet. The black mud on the floor of the sea contains purple pigments formed by a special bacterium, *Thiocapsa roseopersicina*, which lives by "anoxygenic photosynthesis" (in the absence of oxygen and yet requiring sunlight). The changing concentrations of this pigment in sediment cores indicate periods of substantial movement both up and down of the top of the poisonous water mass in the past. Major storms or tsunami generated by earthquakes would have been capable of momentarily *sliding* away the protecting surface layer to release a lethal cloud of destruction across the basin and its rim.

Warned that to attempt a direct crossing of the deadly sea would be

The hero Gilgamesh portrayed in Akkadian bas-relief sculpture

futile, Gilgamesh is directed by a barmaid to a boatman named Urshanabi. However, Gilgamesh is mystified when shown the ferryman's means of propulsion of his craft and in a fit of rage destroys them. "The Stone Things, O Gilgamesh, were what carried me across, that I might not touch the waters of death." All translators have had difficulty with the interpretation of what these "stone things" were. They exist in the context of "senders . . . without which there is no crossing death's waters." Some scholars have thought them to be passive devices such as "talismans," images, or even "lode-stones," though they are clearly in the context of "bearing him along." In fact, after Gilgamesh breaks the "stone things" with his ax, he is admonished by Urshanabi. "It is your hands, Gilgamesh, that prevent the crossing! You have smashed the stone things, you have pulled out their retaining ropes."

The word translated as rope is *urnu* in the original cuneiform inscriptions. Its meaning has also generated much uncertainty. Most commonly the trans-lation is *urnu*-snake, implying that it was the protective guardian of the things of stone. One expert in Oriental studies has noted that *urnu* might be connected with the Egyptian word *Urnes,* the name of a portion of the river in the Egyptian underworld. Others associate *urnu* with vines, since Gilgamesh is instructed to go in the forest and cut more of it after he has slashed the originals with his dagger.

For perhaps thousands of years fishermen and others wishing to go north-ward in the Bosporus against the southward-flowing surface current have simply lowered rocks in a net into the bottom current to be towed north. By the appropriate manipulation of one or more baskets it is possible to hold a boat against the outflow and maneuver across the strait, thereby ferrying passengers and goods to the other side. It is easy to imagine that the "stone things," mysterious even to the guslar reciting the poem, refers to those weighted baskets the ferrymen might have used to maneuver across the Bosporus.

Gilgamesh's ultimate goal had been to meet Utnapishtim and learn about eternal life. He does manage to get across the strait. But what he learns from Utnapishtim is that "there is no word of advice." With compassion Utnapishtim explains the unbearable truth—there is no eternity. Rhetori-cally, he queries, "Do we build houses forever? Do we seal a contract for all time?" Then assertively he continues, "As for death, its time is hidden. The time of life is shown plain."

At this point Utnapishtim makes known to Gilgamesh two closely guarded secrets of the gods. The first is the whole flood narrative of how he had been warned by Ea to build an ark and load it with the seeds of all living things because it was the intent of the gods to destroy all of human-kind by a flood. Ea gave no reason for this horrifying decision of the gods. Utnapishtim obeyed Ea's commands, and when the flood came with terrible violence he and his family floated away for days to finally run aground in the midst of a vast watery expanse. He had sent out birds in search of land,

and when the waters finally receded, they all exited the ark. Utnapishtim prepared a feast for the gods, which they, wretched with hunger, consumed with ravenous delight and was granted immortality and sent to live "at the source of all rivers."

The second is the existence of a plant of rejuvenation, a thorny sagebrush still on the floor of the glistening sea. Utnapishtim instructs, "If you get your hands on that plant (and eat it), you'll have everlasting life."

Gilgamesh accepts the challenge, ties rocks to his feet, dives in the sea, and retrieves the plant. However, on the journey home to Uruk, as Gilgamesh is bathing in a pool of water, a snake catches the sweet fragrance of its flower and steals it away. Weeping in anguish, Gilgamesh understands at last that immortality is not to be his ultimate reward.

The ultimate failure of his quest is a recurring theme not only of the *Gilgamesh* epic but also of most Mesopotamian myths and those subsequently spawned in the literature of ancient Greece. However, the Sumerians and Akkadians, and even the Greeks, did not believe in a reward after death. Death might be postponed through a petition to a god, but no one could escape it. The body returned to clay, and a duplicate "phantom" entered a new abode through an aperture in the grave, leading to an immense, dark, silent, and sad netherworld where one had a torpid and gloomy existence forever.

THERE is an older Sumerian version of the flood simply called "The Deluge." It is short but also very fragmentary. It, too, begins with a story of creation of man and all animal life, the kingships, the five great antediluvian cities, each under the guardianship of a god. Then it is revealed that a number of the gods are bitter and angry because they have found that there has been a decision to destroy all mankind by a flood. No reason is given for the imposition of this disaster. Ziusudra, who is the Sumerian equivalent of Utnapishtim and Noah, is a pious king who believes in divine revelation. One day near a wall he hears a voice warning him that a deluge is being sent to destroy all mankind. He receives instructions. At this point a large part of the tablet is missing. When the story resumes, the flood is in full fury. It also rages for seven days and nights:

> The flood had swept over the land
> And the huge boat had been tossed about by the

windstorm on the great waters,
Utu came forth, who sheds light on heaven [and] earth.

This sun god Utu then "brought his rays into the giant boat." Ziusudra makes a sacrifice: He kills an ox, slaughters a sheep, and prostrates himself before An and Enlil. He is given "life like that of a god" and sent to live in a mythical paradise.

IN most Mesopotamian mythology the purpose of the flood story seems mainly to report a disastrous event, and while each version is colored somewhat by local tradition, in every instance the flood was clearly perceived as a traumatic divide in human history. The flood itself seems to be accepted as a unique natural event caused by godly caprice. The rationale of the gods for inflicting such terror seems rather feeble: because people are making too much noise. If there was a political or moral purpose in the framing of most of these earlier versions of the myth, it is obscure to the point of not being recoverable.

However, the much later version of the story of Noah's Flood in the Old Testament (Genesis 6:9–9:17), written no earlier than the ninth century B.C., has a clearer message and yet possesses the same basic framework as its Mesopotamian counterparts: There is a warning by God and instructions to build a boat and take aboard male and female of selected species, which is done. The flood comes as predicted with terrible swift violence, covering all the land. After many days, calm returns, and finally the ship runs aground. Birds are released with the hope they will find dry land; one does, and slowly the land emerges from the sea. The patriarch builds an altar and makes a burnt offering of every clean beast and every clean fowl.

But the biblical flood story makes one radical departure from the older versions. To begin with, the writers of Genesis appear to be straining to contain two very different ideologies: the old pantheism and the new monotheism. It is the role assigned to the singular God that sets the biblical flood story apart from all its predecessors and that changes the focus of the original history. The central Judaic religious ideology around which the Genesis flood story is woven is that there is only one God, who is almighty, purposeful, and good, and has a direct relationship with the one good man, Noah, the patriarch and progenitor of the generations that survive the flood. As in the earlier versions, the flood is visited on the people as a punishment, but their transgression in Genesis is distinctly a moral one. They are corrupt. God decides that only Noah and his family and selected animals and birds and plants should survive. Furthermore, he needs only one event to accomplish his task. "For my part, I am going to bring a flood of waters on the earth, to destroy under heaven all flesh in which there is a breath of life; everything that is on earth shall die." This is an omnipotent God.

Not all of the more human qualities of the Mesopotamian gods were lost in the Genesis rendering. Enlil and Enki and the rest of the pantheon, dependent as they were upon humans to provide them with food and drink, were desperately hungry after the flood, and when they smelled the sweet savor of the banquet that the Mesopotamian hero had prepared, they gathered around it "like flies." So in Genesis, too, the Lord "smelled a sweet savour" of Noah's burnt offering. Enlil, perhaps feeling contrite, granted Utnapishtim and his wife the immortality of the gods. The Old Testament God, like the Mesopotamian gods, also seems to be a bit taken aback by the enormity of his deed and not only allows Noah to live for a very long time, but makes a covenant never to inflict such a flood on humankind again.

WHILE impatiently awaiting permission in Istanbul to dig at Nineveh, George Smith telegraphed the *London Illustrated News,* promising that he would return from Mesopotamia with a full historical account of the flood. He had gone from believing the flood was an act of God, by whose decree Earth had been inundated and all life destroyed except those on board the ark, to thinking it may have been a natural event, not worldwide but affecting only the peoples of Mesopotamia. In making this commitment, might Smith have supposed that he would discover among the ruins of King Assurbanipal's palace the physical evidence of the flood? Or did he anticipate that he would find among the thousands of tablets still there in the royal library a more expansive account of the deluge, written in the form of a historical document rather than as part of an epic? Unfortunately, this will never be known. In perishing in 1875 from a type of virus that lurks in crowded, disease-infested villages along the sluggish wadis of the intensely hot dry desert, Smith was unable to fulfill his pledge.

Since Smith's death, the search in Mesopotamia for the signs of a large-scale catastrophe has suffered its ups and downs. A bright moment came in 1928 when Leonard Woolley chanced upon a thick layer of homogeneous silt in the ruins of Ur such as would have been laid down by an overbank spill of the nearby Euphrates River. Not long after, another deposit was uncovered upstream in the excavations of Shuruppak, the ancient city mentioned by the poet Sîn-leqi-unninnï in his Babylonian version of *Gilgamesh.* The reporting of these observations in the popular press stimulated the public's imagination across Europe and North America until the much awaited confirmation failed to trace the deposits laterally for any substantial distance—indeed, even from trench to trench within a single archaeological site. Accordingly, the engaging idea that a single grand deluge had engulfed all of southern Mesopotamia fell from favor. What remained in its place, except for those who read the Bible as literally true, was the raw myth alone, its inspiration attributed either to an amalgamation of numerous, unpredictable, and sometimes devastating spring floodings of the Tigris and Euphrates rivers during the snowmelt in the Taurus Mountains, or to a figment of the human imagination distilled into a remarkably uniform account by the smoothing action of retelling by a hundred generations of guslars.

To Ryan and Pitman several key elements of the basic theme of the Mesopotamian flood stories do not make sense in the context of a river flood. For example, there is the warning of the disaster to come, in sufficient time to build and load a boat. River floods in the lower reaches of almost all meandering river valleys, fed by waters that have gathered in distant mountains due to torrential rains or the rapid melt of a winter's accumulated snow, cascade downstream without notice, spilling over the banks, piercing the natural levees, and inundating thousands of acres of land. The annual floods of the Nile fed directly by the monsoon rains in the mountains far to the south are not preceded by any natural alarm. But they occur each year with such regularity as to be fairly predictable. In Mesopotamia, however, the river floods are quite erratic, indirectly sourced as they are by the melting of the snows in the mountains of Anatolia and Armenia. Sometimes the seasonal snows are heavy, sometimes light; sometimes spring arrives early, sometimes late. The magnitude of the rise of the river waters and their time of arrival vary considerably and are unpredictable.

As in the traditional flood myths, the flooding of the Black Sea was replete with vivid warnings. Sea level had been rising inexorably for thousands of years and was still climbing steadily—over eight inches in a human lifetime —when the ocean crested the Bosporus divide. The rise of these ocean waters on the Marmara side must have given some a sense of foreboding, particularly in the last days before the final breaching when, driven by the occasional storm or a strong southerly wind, the waters would have lapped over the top of the earthen dam. One could stand at the top of the sill with one's feet in the sands made damp by this first wash of the ocean waters and look to the north downward through the dry valley to the shimmering distant lake over four hundred feet below. Then there was the first continuous trickle, a thin ribbon of water threading its way through the dirt, leaves, and debris, cutting deep into the soil, growing to a wild tumultuous flume in a matter of days.

By then this flume already bore two hundred times the volume of water that today flows over Niagara Falls, enough to raise the level of the Black Sea six inches every day, filling in yet another mile of the shallow meandering river valleys and their deltas. Only half a mile across at its narrowest point in the Bosporus valley, with the ocean's limitless power behind it, the salt water roared through the strait at speeds greater than fifty miles an hour, crashing through unabated, radiating a thunderous din and vibration that could probably have been heard and felt around the entire rim of the Black Sea. Although this was the final warning, there was still time to flee.

In all versions of the flood myth the warning and the deliberate instructions for building and loading the ark suggest that this would be a permanent flooding. "Dismantle your house, build a boat. Reject possessions, and save living things," and "I loaded her with everything there . . . loaded her with all the seed of living things, all of them." As the days stretched to weeks and months, and the flood waters continued to rise, these Black Sea people must have realized that this would be a permanent flooding. They would never see their homes again. They would have to abandon their ancestral land, never to return.

But floods of alluvial plains are rarely, if ever, permanent. People flee, wait out the flood, and return to restore their homes and fields. Such experiences in Mesopotamia may well have served to reinforce the myth memory of a permanent flood, and the lack of such reinforcement may account for the puzzling absence of a flood mythology in Europe. Many great rivers flood catastrophically, but these events rarely give rise to mythology.

THE Bosporus flume roared and surged at full spate for at least three hundred days. Some had probably been there to witness its violent birth. Others, very courageous indeed, were drawn to see its monstrous flow. Anyone who saw it must have carried away frightening images of the fury and power of the sinuous jet of salt water. The flume may have seemed like a huge, endless serpent writhing through the narrow defile, continuously roaring, sending up billows of mist like smoke from its fiery mouth, destroying all in its path, and pouring the alien salt water of death into the sweet water below. Eventually, for no visible cause, the flow through the Bosporus slowed, and the level of the Black Sea surface reached that of the Aegean. The wild ocean had been subdued by forces or gods unknown. The salt water from the ocean beyond had mixed chaotically with the fresh of the Black Sea, but all was quiet now. Beaches and dunes formed and ringed the now tranquil water like a fence. But it would be a long time before humans would again settle along its shore.

The mythology of the Middle East, particularly those legends concerned with creation, is sprinkled with images that seem to commemorate these events. The mythologists Robert Graves and Raphael Patai gleaned from fragments in the Bible and other ancient Hebraic writings numerous images that seem to do so. In one, "The roaring waters of the Deep arose and Tehom their Queen threatened to flood God's handiwork." But God rode out on his fiery chariot, and dispatched her allies Leviathan and Rahab with club and sword, cowed Tiamat into submissiveness and confined her to live within the boundary of the sand dunes, "making a decree which Tehom could never break however violently her salt waves might rage, she being, as it were, locked behind gates across which a bolt has been shot." Tehom is said to be the dreaded Babylonian goddess Tiamat. And in another image of one of Tehom's pets, "Leviathan's monstrous tusks spread terror, from his mouth issued fire and flame, from his nostrils smoke, from his eyes a fierce beam of light; his heart was without pity. He roamed at will on the surface of the sea, leaving a resplendent wake; or through its lowest abyss making it boil like a pot."

The ancient Greeks had a number of flood myths. The Deucalion flood myth, which has the same basic story as those of Mesopotamia, is presumed to have been passed on to them by the Phoenicians. But there is one tale from the island of Samothrace, which lies in the Aegean just to the west of the entrance to the Dardanelles, which may describe the actual events of the Black Sea flood—in reverse. The only source for this story is from the writings of Diodorus of Sicily, who passed on a summary. "The first and original inhabitants used an ancient language which was peculiar to them and of which many words are preserved to this day in the ritual of their sacrifices." According to Diodorus, Euxinos Pontos (the ancient Greek name for the Black Sea), which at the time was a lake, became so swollen by the waters in the rivers that flowed into it that its waters burst forth violently through a natural earthen dam out through the outlet at the Cyanean Rocks (located at the Black Sea end of the Bosporus) and through the Hellespont. This flood inundated a large part of the coast of Asia Minor and made no small amount of the level part of the land of Samothrace into a sea. And this is the reason, we are told, why in later times fishermen have now and then brought up in their nets the stone capitols of columns, since even cities were covered by the inundation.

But this reverse flood, a Black Sea lake filled with water to a level well above the Aegean, bursting through into the Bosporus and flooding the island of Samothrace, is not plausible. The Aegean is connected to the world ocean through the Mediterranean and the straits of Gibraltar, so that any water flowing from the Black Sea through the Bosporus and the Dardanelles to the Aegean would have been rapidly spread out across this vast surface, causing no significant rise in the level of the Aegean. It seems more likely that the ancestors of these people had originally lived by the Black Sea lake shore or in the Bosporus Valley itself and had witnessed the breaching of the Bosporus dam and the rush of water from the ocean into the Black Sea. Their fields and homes were inundated, and they fled, eventually settling on Samothrace, perhaps becoming integrated with the native inhabitants. Like almost all people with strong oral traditions, whose history is recorded in myths, they would come to identify their origins with this island, claiming they had always lived there, that they had "sprung from the rock." So the legend of their flood, accommodated to their presumed island origin, was inverted, and the water instead of flowing into the Black Sea, was somehow remembered by them to have flowed outward, flooding their Samothrace home. But the specificity of the detail, such as the detail of the earthen dam, is tantalizing.

T H E flood myth lives for a number of reasons. First, it is surely a true story of the permanent destruction of a land and its people and a culture suddenly and catastrophically inundated, of farmers uprooted from their hard-won fields, their villages permanently destroyed. They had to flee with what little they could carry, old and young, straggling along day after day. They had to flee and at the same time obtain food from the land by hunting and gathering, skills they had long forgotten. Some, perhaps many, probably died of exhaustion. For those who had lived near the Bosporus at the western end of the lake, the sight and sound of the flume must have filled their hearts with terror and horror, "like the bellowing of a bull, like a wild ass screaming. . . . Earth shook, her foundations trembled, the sun darkened, lightning flashed, thunder pealed, and a deafening voice, the like of which was never heard before, rolled across mountain and plain." So the tragedy was indelibly implanted in their oral history.

Over the thousands of years since, with war, invasion, migrations, and other calamities, the legend disappeared from the folk memory of many. However, to those who fled to Mesopotamia and whose progeny are still there, the flood lived on for thousands of years, its telling and memory

A *Telling of* Atrahasis

IN the twelfth year of the reign of King Ammi-saduqa of Assyria and in accordance with the law of Hammurabi, the caravan makes its encampment outside the fortified walls of the ancient city, on the greensward by the river. The voyagers bathe themselves in its muddy waters and buy fresh food in the market, for this is to be a night of feasting. The journey upriver from Babylon was arduous due to unusually early spring floods, which forced them to travel along the west bank of the Euphrates where the ground was higher but the danger from the barbarian raiders of the desert was much greater. Nothing untoward occurred except that one of their guard dogs was snatched away by a pack of lions.

After a much needed rest and some trading in the bazaar, this caravan will attach to another convoy and proceed north through the land of Hana to Emar and thence across the dry steppe to Yamhad. For many that infamous trading post is their final destination, but for the remainder the route continues to the Kebir River and on to the port city of Ugarit along the coast of the azure sea littered with its coveted shells of cowry and conch.

Here on the outskirts of Mari, the Euphrates has cut into the nodular yellow chalk of the opposing bluffs a broad valley over three leagues wide, through which the strong river meanders on its journey south to Sumer and from there into the vast reed marsh before arriving at the salty gulf from which comes the copper and ivory of Tilmun. In many places the Euphrates valley is a productive green carpet nurtured by the sporadic seasonal rains and the spring overflow of the river. The yearly floods are as unpredictable as their source—the melting of the deep snow cover and alpine glaciers in the tall mountains around Urartu and beyond. Along the lush riverbanks date palms grow in parallel rows; in their shade, pomegranate, citrus, and

*Nur-Aya, the renowned scribe and storyteller, recites the flood story
of Atrahasis circa the mid-eighteenth century B.C.*

apple release buds above a ground cover of herbs and spices. Meticulously
tilled fields of wheat and barley alternate with gardens of garlic, onion,
lettuce, cucumber, and lentils, irrigated through a network of canals by the

constant workings of the waterwheel and the shadoof. Outside this narrow farm belt, as far as the eye can see in the direction of the setting sun, rolling hills stretch away to a barren landscape too dry for human settlement, while in the opposite direction the Jazirah Desert nearly fills the huge expanse between the two great rivers.

There are several caravans camped in this verdant park. As is the custom in the evening, some of the younger men and boys and their dogs remain near the tents to look after the donkeys and goats while the others share food, wine, and beer around a fire. This particular evening the fireside is crowded. Nur-Aya, the renowned scribe and storyteller, has come up from Sippar to entertain the royal court in the great palace built by Zimri-Lim almost two hundred years earlier. Tonight Nur-Aya is out among the caravan folk to be inspired by their chatter and personal stories. He has spent his life as a bard, learning and spinning tales in and around all the great cities of Elam, Babylonia, and Akkad. Accompanied by servants and guards, courtesy of the governor, he is on his way to live out the remainder of his days in Ugarit. On each of the many preceding nights he has enthralled other so-journers with his songs and poems. Some in this group will break away in the morning to the oasis of Tadmor on the track to Damascus, but they all, and particularly the children, plead with him to tell the tale of the Great Flood and of Atrahasis, its hero.

The night is clear. A multitude of stars streams across its firmament. The river remains flush with meltwater, its load of silt brought to rejuvenate the fields. Date palms wave slowly in a breeze that carries the sweet smell of the newly opened blossoms from the orchards and gardens. The zephyrs caress the cheeks of excited children. Once everyone settles around the burning cinders—the young snuggling with their mothers, wrapped in shawls to ward off the chill—Nur-Aya rises to his feet. A hush falls upon the crowd.

The poet begins to chant slowly in a deep resonant voice that travels easily to the edge of the river. To the accompaniment of an eleven-string lyre he sings of the time near the beginning when the gods cast the lots and made the division, when Anu goes up to the sky and Enlil takes the Earth for his people. The bolt that bars the sea is assigned to the farsighted and clever Enki. The Anunnaki of the sky order the Igigi to carry the workload for all the gods. When the Igigi revolt, people are created and put on the Earth as slaves to provide labor for the gods.

The lines of his poem give new vigor to the timeless legend brought to ancient Uruk by the seven sages of the distant past. Nur-Aya uses the fixed meters of his lines as guideposts, his fertile imagination masterfully weaving the thread of a thrilling tapestry full of rich details to add texture and content to the well-known plot of the legend. He recounts the horrors of plagues and relentless famines when people were driven to desperation and practiced cannibalism.

The breeze rustles the palms and ripples the flames as the old man chants, his audience repeating each refrain like a well-rehearsed chorus:

> Six hundred years, less than six hundred, passed
> And the country became too wide, the people too numerous.
> The country was as noisy as a bellowing bull.
> The God grew restless at their racket.

The story continues. The boisterous noise of an overpopulated land is disturbing Enlil's sleep. He decides to wipe them out, all of them, all of humankind, with a great flood.

There is a good and especially wise man in one of the towns, Atrahasis, son of Ubara-Tutu, whose ear is open to his god, Enki. However, Enki is required to swear an oath in the midst of the other gods that he will not tell the people of the coming doom. Then he sees a way out of his dilemma. He can warn of the flood by speaking to the wall of Atrahasis's house.

> Reed hut, make sure you attend to all my words
> Dismantle your house, build a boat.
> Reject possessions, and save living things.

The world is to be destroyed; there will be no returning to this place. Atrahasis is to take with him all things necessary to start life anew elsewhere. He is even to deceive his own people by not informing them of the purpose of the boat or caution them of Enlil's displeasure. All Atrahasis can say is:

> My god is out of favor with your god.
> Enki and Enlil have become angry with each other.
> They have driven me out of my house.
> Since I always stand in awe of Enki,
> He told me of this matter.
> I can no longer stay at home.

Atrahasis employs carpenters, reed workers, and those who use bitumen to seal cracks to build his boat and seal its cracks. No one knows its real purpose.

As the campfire burns, its flames lighting up faces in the crowd, the bard elaborates a full description of the boat, its immense size, the food and beer being consumed by the workmen, how Atrahasis selects all kinds of animals, birds, and seed to put onboard, and finally his family, too. But only he, Atrahasis, knows what is to come. While his family is eating, drinking, and merrymaking, not suspecting a violent catastrophe is to come, Atrahasis becomes agitated.

> He went in and out,
> Could not stay still or rest on his haunches,
> His heart was breaking and he was vomiting bile.
> The face of the weather changed.
> Adad bellowed from the clouds.

The voice of Adad, the storm god, peals across the land, louder and more fearsome than any thunder ever heard. It is the warning signal that the flood is near at hand. Atrahasis seals the door of his ship, and with the wind raging through its rigging, he goes up on deck to cut loose the mooring ropes, casting his ship adrift. The flood arrives in all its fury:

> The Flood roared like a bull,
> Like a wild ass screaming the winds howled.
> The darkness was total, there was no sun.

Nur-Aya tells of the fear that grips everyone, of how the flood weapon sweeps in among the people like an army. It tears out the mooring poles of their boat. It bursts through the dikes and submerges the fields, sweeping away all life before it. The bolt that bars the sea is opened. The ground trembles. A blinding mist whirls in upon them, shrouding them in darkness. Even the Igigi, the lesser gods, are terrified. They flee up to the heaven of Anu, where they cower like dogs.

Nur-Aya explains how the boat is lifted by the rising water and, driven by the wind, is carried out to sea; how Atrahasis and his family can barely see the hills of their homeland through the mist as it fades and sinks into the abyss.

The greater gods, the Anunnaki, huddle terrified in Anu's heaven. They hear Nintu, the matriarch and mother of all the people, shriek in agony at what has been wrought, at the sight of her dead offspring floating on the water like sheep that have been sacrificed.

> She was sated with grief, she longed for beer in vain.
> Where she sat weeping there the great gods sat too,
> But like sheep, could only fill their windpipes with bleating.
> Thirsty as they were, their lips
> Discharged only the rime of famine.
> For seven days and seven nights
> The torrent, storm, and flood came on.

The transfixed crowd at Nur-Aya's feet hears how Atrahasis looks out when the storm abates, only to find nothing but water in every direction, an unending rippled flatness that shimmers in the light, unbroken by even the smallest object. It is as if the sea has risen above the highest peaks of the mountains. There is no sound but the lapping of the water against the boat and the bleating of the animals below decks. They float for days across the endless expanse, seeing neither land nor life, until finally in desperation Atrahasis releases some birds. These are not diving seabirds but land birds that will have to find a firm footing to roost. When set free they circle the ship, spiraling high above in ever-widening circles until they fly out of sight, only to return by dusk. They have not sighted a landfall. In the following days Atrahasis releases other birds, and again all come back to the ship. But the third time they disappear for good. The flood survivor rejoices. Land cannot be far away.

The dark blue-gray of a line of hills stands low along the horizon, silhouetted against the early light. As the ship is propelled toward this wonder, the low hills merge into one and seem to rise up out of the water and become a mountain that looms above them. At last they finally land on its strange shore.

Atrahasis opens the hatches of his vessel and sends out all the living creatures. He makes a sacrifice as an offering to the gods. He prepares a succulent feast of lamb cooked in savory herbs and spices, with breads, dates and figs, sweetmeats, wine and beer. The gods, with the rime of famine on their lips, descend from the safety of their heavenly perch. Gathered like flies over the offering, ravenous with hunger and parched for drink, they gorge themselves until all is consumed.

Nur-Aya tells his rapt audience how angry and embittered the gods are with one another, and particularly with Enlil. Because of him all life has been extinguished from the face of the Earth. At last Enlil expresses his regret. He sets the course to repopulate the land. Enlil blesses Atrahasis and his wife, granting them immortality and sending them to live in a distant land.

It is late. The aging guslar has held the crowd entranced. Only a few of the children have fallen asleep. The swollen river sings, too, as it rushes downstream with its waters and mud that will once again refresh the soils of the alluvial plain.

Nur-Aya finishes with the last lines of the timeless story, reciting words that announce not only an end but a beginning.

> I shall sing of the flood to all people:
> Listen!

Acknowledgments

I T has been more than twenty-five years since a colleague, John Dewey, in an off-hand remark, suggested that Noah's Flood might be found among the inundated and still-flooded basins of the Middle East. However, it wasn't until 1991 that we began serious work to follow up on his proposition. During the very earliest years, Millie Alvarez assisted us by tracking down the many relevant flood legends, their various translations, and their interpretations.

As our research efforts intensified, we were given invaluable assistance by Ashley Ryan Gaddis, Jane Haxby, and Amanda Pitman, all of whom diligently scoured libraries, helping to research various chapters of our manuscript. Mary Ann Brueckner and Susan Klimley tracked down an endless stream of obscure books and documents while Bel Hautau helped with organizing the research materials and assembling a reference database.

In the initial stages we received much needed advice from Professor Ralph Solecki of Columbia University. More recently, discussions with a number of other archaeologists have been vital to our synthesis. These include professors Ian Hodder, David Harris, Cyprian Broodbank, Colin Renfrew, Jane Renfrew, Joan Oates, Andrew Moore, Andrew Sherratt, Peter Bugucki, Douglas Bailey, Neil Roberts, Fekri Hassan, Ofer Bar Yosef, and James Mellaart.

We were very fortunate that our ideas caught the attention of the British Broadcasting Corporation. Their production of the documentary "Noah's Flood" in 1996 provided us with additional stimulating contacts both while the film was in the making and after it was aired. We are grateful to John Lynch of BBC Horizon, to Antonia Benedek and David Collison of Third Eye Productions, and to Richard Curson-Smith, the film's director, and to the crew, for this occasion.

We are deeply indebted to John Dewey, who sowed the seed of our search. Captain Hüseyin Yüce, Naci Görür, Mehmet Sakinç, and Petko Dimitrov provided critical data on the Turkish Straits and the Black Sea continental shelf. Çelal Sengor enthusiastically helped us all along the way. He opened many doors for us in Turkey, allowed the unlimited use of his library, and gave fully of his encyclopedic mind. Martine Rossignol-Strick unearthed critical scientific literature at just the right time.

Our exciting new data materialized during a scientific expedition on the Black Sea aboard the Russian research vessel *Aquanaut* arranged by Evgeny Kontar and Ruben Kos'yan of the P. P. Shirshov Institution of Oceanology. The splendid success of this adventure was due in no small part to the vast experience of mission chief scientist Kazimieras Shimkus and to the stalwart efforts of Candace Major, our technician and now a colleague. We are also deeply indebted to the officers and crew of the *Aquanaut*, who cheerfully assisted in every possible way.

Recorded interviews provided interesting insights and fresh information. We would like to thank Glenn Jones, George Kukla, Richard Fairbanks, David Ross, Nicolae Panin, Sigfus and Pauline Johnsen, and James Mellaart for tolerating the intrusion and for the many remarkable tales we were told.

We thank Bill Haxby for his digitally generated maps and Natasa Sotiropoulos for her splendid and imaginative drawings.

A number of friends and colleagues read chapters of the manuscript during their development and made important suggestions. Among them are Charlotte Schreiber, Faye and Stan Yates, Kim Kastens, Linda Punderson, Janet Fox, Lincoln Pratson, Susan Moeller, and Natasa Sotiropoulos.

We thank Bob Asahina, formerly of Simon & Schuster, our present editor, Bob Bender of Simon & Schuster, and Leonard Mayhew, all of whom constantly strove to make this a better book. We are particularly grateful to our agent Roger Jellinek, without whose encouragement and effort at the very formative stages this project would never have gotten off the ground and whose constant guidance through the writing process has been invaluable.

WP is especially appreciative of his daughters Cordelia and Amanda and WR of his daughters Sarah and Ashley, son Sean, and wife Judy for their unflagging interest, help in research and editing, as well as patience. Lastly we owe much to the institution within which we both have worked for over thirty years—when we began there, it was called Lamont Geological Observatory; now it is Lamont-Doherty Earth Observatory of Columbia University.

Notes

Page

13 *up the mighty river:* describes the ancient Danube River draining vast regions of eastern Europe through a narrow gap between the Transylvanian Alps of Romania and the Balkan Mountains of Bulgaria and Yugoslavia.

13 *the ruins of an abandoned shrine:* refers to Lepenski Vir, a prehistoric human settlement on the Serbian side of the Iron Gates Gorge of the Danube River, discovered in 1967 by Dragoslav Srejovic and containing the most sophisticated Stone Age (Mesolithic) foraging culture of Europe. It reached its zenith during early sixth millennium B.C. with the world's first monumental sculpture and then quite suddenly collapsed. B. M. Fagan, ed., *The Oxford Companion to Archaeology* (Oxford: Oxford University Press, 1996), 388; D. Srejovic, "The Mesolithic of Serbia and Montenegro," in *The Mesolithic in Europe,* ed. C. Bonsall (Edinburgh: Edinburgh University Press, 1989), 481–91.

13 *a human face:* is reproduced in M. Gimbutas, *The Language of the Goddess* (New York: HarperCollins, 1991), 260; plate 19 (side view) and figure 407 (front view), carved into a boulder of reddish sandstone more than a foot in diameter.

13 *wide lips tightening:* D. Srejovic, "Europe's First Monumental Sculpture: New Discoveries at Lepenski Vir," in *New Aspects of Antiquity,* ed. M. Wheeler (London: Thames and Hudson, 1972).

14 *a few goats:* refers to the Near Eastern species *Capra aegagrus.*

14 *A sizable river:* the ancient Rioni River in Georgia whose tributaries originate in extensive watersheds along the southwestern slope of the Caucasus Mountains.

14 *the lake from which they are fleeing:* refers to the vast New Euxine Lake that occupied the Black Sea region during the last Ice Age and that persisted until mid sixth millennium B.C. (see chapter 9).

15 *houses of reed:* describes a people living in the vicinity of the mouth

of the ancient Dniester and Dnieper rivers where they cut across the now-submerged shelf of Ukraine and entered the New Euxine Lake (former Black Sea) more than one hundred miles offshore of the present coast.

15 *living mussels: Dreissenal polymorpha,* which were prolific in the littoral zone of the New Euxine freshwater lake and its river tributaries. L. A. Nevesskaya, "Late Quaternary Bivalve Mollusks of the Black Sea: Their Systematics and Ecology," *Transactions of the Institute of Paleontology of the USSR Academy of Sciences* 105 (1965): 1–390.

16 *on the delta:* the ancient Sakarya River in northwestern Turkey.

16 *post and beam houses:* an early agricultural village of the Fikirtepe culture in northwestern Turkey during the early sixth millennium B.C. M. Özdogan, "A Surface Survey for Prehistoric and Early Historic Sites in Northwestern Turkey," in *National Geographic Research Reports for 1979* (1985), 517–41; M. Özdagan, "Pendik, a Neolithic Site of Fikirtepe Culture in the Marmara Region," in *Beiträge zur Altertumskunde Kleinasiens, Festschrift für K. Bittel,* eds. R. M. Boehmer and H. Hauptmann (Mainz, Germany: University of Mainz, 1983), 401–11.

16 *another rim of the lake:* along the shoreline of the New Euxine Lake, some fifty miles offshore of the modern city of Burgas, Bulgaria.

17 *From a plateau:* a panoramic view of the Bosporus valley from its European side.

I. DECIPHERING THE LEGEND

Page

21 *his weight tore the brace free:* H. C. Rawlinson, *Archaeologia* 34 (1852): 75.

21 *Behistun Rock:* a tall mountain with twin peaks some twenty-two miles east of Kermanshah, Iran, on the rocky slopes of the Alvend, a segment of the Zagros Mountains that is twelve thousand feet in elevation.

22 *winged deity:* the Ahura Mazda of ancient Persia and the sun god Ashur in older Akkadian carvings. E. Chimera, *They Wrote on Clay* (Chicago: University of Chicago Press, 1938), illustrations on 128–29.

23 *a scholarship to Ealing:* S. Lloyd, *Foundations in the Dust,* 2nd ed. (London: Thames and Hudson, 1947), 13.

23 *to conform the evidences:* W. Buckland, *Reliquiae Diluvianae, or, Observations on the Organic Remains Contained in Caves, Fissures, and Diluvial Gravel, and on Other Geological Phenomena Attesting to the Action of an Universal Deluge* (London: John Murray, 1823).

24 *"a deluge will engulf"*: L. Renou, *Vedic India*, vol. 3 (Delhi: Indological Book House, 1971), 32.

24 *sprung from some common source*: G. Cannon, *The Life and Mind of Oriental Jones: Sir William Jones, the Father of Modern Linguistics* (Cambridge: Cambridge University Press, 1990), 245.

24 *Noah's son Japhet*: In the "Table of Nations" reference in Genesis 10, the various races are portrayed as having sprung from a single stock with an assumed original common language, as pointed out by N. M. Sarna, *Understanding Genesis* (New York: Schocken Books, 1970), 65. The sons of Noah who came out of the ark dispersed in clans to separate lands with their own language and customs.

24 *paymaster to the First Grenadier Regiment*: G. Rawlinson, *Memoir of Major-General Sir Henry Creswicke Rawlinson* (London: 1898).

25 *accompany him on horseback*: S. Lloyd, *Foundations in the Dust*, 77.

25 *May I be no son of Darius*: Herodotus 7–11, written in the fifth century B.C. D. Grene, *Herodotus: The History* (Chicago: University of Chicago Press, 1987), 473.

26 *I Darius, the great King*: E. Doblhofer, *Voices in Stone* (New York: Viking Press, 1961; repr. A. M. Kelley, 1973; trans. by M. Saville), 117.

26 *Ye who in future pass*: Ibid., 37.

26 *"fought its way into the country of myth"*: Thucydides, I.21.8.

26 *lacked the language skills to understand them*: D. Grene, *Herodotus: The History*, 2.

27 *a reed stylus*: C. B. F. Walker, *Cuneiform* (Berkeley: University of California Press, 1987), 22.

27 *when Smith was just a child*: H. McCall, *Mesopotamian Myths* (Austin: University of Texas Press, 1990), 13.

27 *labeled the* mythological and mythical: G. Smith, *Assyrian Discoveries: An Account of Explorations and Discoveries on the Site of Nineveh, During 1873 and 1874* (New York: Scribner, Armstrong & Co., 1875), 4.

27 *obsessed by scripture*: E. Doblhofer, *Voices in Stone*, 134.

28 *On looking down*: G. Smith, *Assyrian Discoveries*, 4.

28 *to tear off his clothes*: H. McCall, *Mesopotamian Myths*, 14.

28 *a work of supposed divine authorship*: G. Smith, *The Chaldean Account of Genesis* (New York: Scribner, Armstrong & Co., 1876), 307.

29 *"a large and distinguished company"*: *The Times*, no. 27,551 (Wednesday, Dec. 4, 1872), 7.

29 *a cast of characters*: In 1872 the phonetic reading of the Babylonian names was very obscure. Smith's initial translation of the name of the

survivor of the flood was Sisit, most likely borrowed from either Sis-
ithrus or Xisuthrus in the flood stories of the Greek poet Berosus. G.
Smith, *The Chaldean Account of Genesis,* 42–47, 51.

30 *authored in the third millennium B.C.: The Times,* no. 27,552 (Thursday,
Dec. 5, 1872), 9.

30 *"it could hardly be doubted": The Times,* no. 27,552 (Thursday, Dec. 5,
1872), 9.

2. Conversions

Page

31 *"ocean of ice":* L. Agassiz, "Discours prononcé à l'ouverture des séances
de la Société Helvétique des Sciences Naturelles à Neuchâtel, le 24
Juillet 1837," *Actes Société Helvétique des Sciences Naturelles* 22 (1837):
v-xxxii, v; J. Marcou, *Life, Letters and Works of Louis Agassiz,* vol. 1
(New York: Macmillan and Co., 1896), 89–108.

31 *as far as the Caspian Sea:* L. Agassiz, "A Period in the History of Our
Planet," *Edinburgh New Philosophical Journal* (1843): 1–29.

31 *the role of mountain glaciers:* J. G. F. de Charpentier, "Notice sur la
cause probable du transport des blocs erratiques de la Suisse," *Ann.
Mines* 8 (1835): 219.

31 pierre à bot: C. Maclaren, "The Glacial Theory of Prof. Agassiz of Neuchâtel," *American Journal of Science* 42 (1842): 358–60.

31 *"the polished, striated, and furrowed surfaces":* W. Buckland, "Memoir
on the Evidences of Glaciers in Scotland and North of England," *Proceedings: Geological Society of London* 3 (1840): 335.

33 *melting icebergs:* C. Lyell, *Principles of Geology: Being an Attempt to
Explain the Former Changes in the Earth's Surface by Reference to
Causes Now in Operation,* 3 vols. (London: Murray, 1830–33).

33 *denuding and harvesting:* G. F. Wright, "Agassiz and the Ice Age," *The
American Naturalists* 32, no. 375 (1897): 167.

33 *unknowingly lowered:* C. F. Holder, *Louis Agassiz: His Life and Work*
(New York: G. P. Putnam's Sons, 1893), 76.

33 *Agassiz was ill prepared:* J. Imbrie and K. P. Imbrie, *Ice Ages: Solving
the Mystery* (Cambridge, Mass.: Harvard University Press, 1986), 26.

33 *In general, I was convinced:* Ibid., 30. Agassiz's daughter wrote that
when her father led the field trip into the Jura, one of the eminent critics,
Leopold von Buch, "could hardly contain his indignation, mingled with
contempt, for what seemed to him the view of a youthful and inexperienced observer." E. C. Agassiz, *Louis Agassiz, His Life and Correspondence,* 2 vols. (Boston: Houghton Mifflin & Co., 1886).

33 *Studies on Glaciers:* L. Agassiz, "On Glaciers and the Evidence of Their Having One Existed in Scotland, Ireland and England," *Proceedings: Geological Society of London* 3 (1840): 327–32.

34 *Blackford Hill:* now known as Agassiz Rock. G. W. White and C. J. Schneer, "Lyell's Life and Works Revisited," *Geotimes*, March 1976: 16.

34 *"That is the work of ice!":* C. Maclaren, "The Glacial Theory of Prof. Agassiz of Neuchâtel," *American Journal of Science* 42 (1842): 351.

3. VISIONS OF PALACES

Page

36 *The Arabian Nights:* A. H. Layard, *Autobiography and Letters,* 2 vols. (London: John Murray, 1903), 26.

36 *through the domain of Bedouin:* S. Lloyd, *Foundations in the Dust,* 2nd ed. (London: Thames and Hudson, 1947), 88.

37 *I look back:* A. H. Layard, *Early Adventures in Persia, Susiana, and Babylonia* (London: John Murray, 1887).

38 *The scene around:* A. H. Layard, *Nineveh and Its Remains,* 2 vols. (London: John Murray, 1849), vol. 1.

38 *My curiosity had been greatly excited:* Ibid., 9.

38 *Visions of palaces:* Layard, quoted in S. Lloyd, *Foundations in the Dust,* 101.

39 *Nimrud, the name given to the great-grandson of Noah:* Genesis 10:8–10. "Noah begat Ham, Ham begat Cush. And Cush begat Nimrod. . . . Out of the land went forth Asshur, and builded Nineveh, and the city Rehoboth, and Calah."

40 *a bizarre anthropomorphic rendition:* A. H. Layard, *Nineveh and Its Remains,* vol. 1, 128–29; A. H. Layard, *The Monuments of Nineveh,* 2 vols. (London: John Murray, 1849), vol. 1, plate 33.

40 *As the sun went down:* A. H. Layard, *Nineveh and Its Remains,* vol. 1, 79–80.

41 *paved with burnt brick:* J. E. Curtis and J. E. Reade, *Art and Empire* (New York: The Metropolitan Museum of New York, 1995), 23.

41 *I built a pillar over against his city gate:* Inscription from Ashurnasirpal's palace at Nimrud translated for the exhibition *Art and Empire: Treasures from Assyria in the British Museum* held at the Metropolitan Museum of Art, New York, May 2–August 13, 1995.

42 *"he had more than once":* E. Doblhofer, *Voices in Stone.* (New York: Viking Press, 1961; repr. A. M. Kelley, 1973; trans. by M. Savil), 126.

43 *In this magnificent edifice:* A. H. Layard, *Discoveries in the Ruins of Nineveh and Babylon* (London: John Murray, 1853), 589.

43 *with chain gangs:* A. H. Layard, *A Second Series of the Monuments of Nineveh,* 2 vols. (London: John Murray, 1853), vol. 2, plate 15.

44 *Assyrian revival became the fashion of the day:* H. McCall, *Mesopotamian Myths* (Austin: University of Texas Press, 1990), 12.

4 . THE FACE OF THE DEEP

Page

45 *"It appeared likely that a search":* G. Smith, *The Chaldean Account of Genesis* (New York: Scribner, Armstrong & Co., 1876), 3.

46 *"several coincidences between the geography":* Ibid.

46 *"the ship rested on the mountains of Nizir":* Ibid., 4.

46 *This search was a long:* Ibid., 5.

47 *"poetical and exaggerated":* G. Smith, *Assyrian Discoveries: An Account of Explorations and Discoveries on the Site of Nineveh, During 1873 and 1874* (New York: Scribner, Armstrong & Co., 1875), 222.

47 *"a basis of truth":* Ibid.

47 *"bring back the lost tablets":* Ibid., 14.

47 *I found a new fragment:* Ibid., 95.

48 *contained the greater portion:* Ibid., 97.

48 *"the tradition of a seafaring people":* Ibid., 214.

48 *"a national poem":* Ibid., 205.

48 *"the waters of the fountain":* Ibid., 168.

48 *was a fragment evidently belonging:* G. Smith, *The Chaldean Account of Genesis,* 7.

48 *"one giving the Creation and fall of man":* Ibid.

49 *I [have] excavated: Daily Telegraph,* March 4, 1875.

50 *When above, were not raised:* G. Smith, *The Chaldean Account of Genesis,* 62.

50 *a bolt shot across a gate:* R. Graves and R. Patai, *Hebrew Myths* (New York: Doubleday, 1989), 29.

50 *"guarded by a sword":* G. Smith, *The Chaldean Account of Genesis,* 88.

51 *when my investigations are completed:* H. McCall, *Mesopotamian Myths* (Austin: University of Texas Press, 1990), 14.

5 . UR OF THE CHALDEES

Page

52 *"Kings-Lists":* C. L. Woolley, *The Sumerians* (New York: W. W. Norton, 1965), 21–26; G. Roux, *Ancient Iraq,* 2nd ed. (London: Penguin Books, 1980), 41; J. Oates, *Babylon,* rev. ed. (London: Thames and Hudson, 1986), 18, 22–23.

52 *a tablet:* G. Smith, *The Chaldean Account of Genesis* (New York: Scribner, Armstrong & Co., 1876), 264. His translation of what we now call the *Epic of Gilgamesh,* tablet 11, col. I, lines 11–12, gives the name of the city Shrippak.

52 *"Man of Shuruppak":* J. B. Pritchard, *The Ancient Near East,* 6th ed., 2 vols. (Princeton: Princeton University Press, 1973), vol. 1, 66; provides one form of the more common spellings and assigns the Euphrates as the river. Shuruppak also appears in an older text of clear Sumerian authorship in a myth entitled *Deluge,* where it is mentioned in line 96 (translated by S. N Kramer) as one of the five cult-centers before the flood.

52 *"tear down his house":* Ibid., 66.

52 *"Euphrates' banks":* Ibid.

52 *about halfway between Babylon:* Over the span of millennia the two great Mesopotamia streams wandered back and forth across the floodplain. The deltas at the rivermouths filled in a large coastal lagoon inherited from the Ice Age, turning the sea to marsh, then to arable land, and thereby expanding the region that would become known as Sumer. Inevitably the maritime ports of the first settlers that had been built on riverbanks not far from the sea became abandoned to inland desert, isolated by the natural evolution of the landscape from the very arteries that had nurtured their commerce and watered their farmlands.

52 *assistant director:* B. M. Fagan, ed., *The Oxford Companion to Archaeology* (Oxford: Oxford University Press, 1996), 761.

53 *a vivid picture of Old Babylonian life:* J. Oates, *Babylon,* rev. ed. (London: Thames and Hudson, 1986), 76–77.

53 *Broad Street, Church Lane, and Paternoster Row:* Ibid., 77.

53 *life span of hundreds of years:* The reigns of the ten antediluvian kings total an unlikely quarter of a million years!

53 *one of these monarchs:* For example, at the mound of al-'Ubaid near Ur, Woolley found a block of limestone with an inscription on it announcing a dedication by A-anni-padda, king of Ur, son of Mes-annni-padda, king of Ur. The latter name appeared on the king-lists as founder of the First Dynasty after the flood.

54 *"Are the king-lists":* C. L. Woolley, *The Royal Cemetery,* 10 vols. *Excavations at Ur* (London: 1934), vol. 2, 30.

54 *"We need not try":* Ibid., 31.

54 *"The total destruction of the human race":* Ibid., 32.

54 *human sacrifice on an extravagant scale:* Ibid., 39.

54 *The entire retinue of the ancient court:* see B. M. Fagan, *Eyewitness to Discovery* (Oxford: Oxford University Press, 1996), 132–40, for an

illuminating reprint of the royal grave discoveries in Woolley's own words.

54 *the very first settlers:* called the Ubaids after the Tell al-Ubaid located four miles west of Ur; opened toward the end of Woolley's excavations and containing an especially rich assemblage of their unique pottery and flint tools; H. R. Hall, *A Season's Work at Ur, Al-'Ubaid, Abu Shahrain (Eridu), and Elsewhere.* (London: Methuen, 1930), 229.

54 *little more than a marsh village:* B. M. Fagan, ed., *The Oxford Companion to Archaeology,* 730.

54 *the material evidence of a great inundation:* J. Hawkes and Sir L. Woolley, *Prehistory and the Beginnings of Civilization,* vol. I, *History of Mankind* (New York: Harper & Row, 1963), 368.

54 *it was easy to believe:* Ibid.

55 *"With the land too valuable to be left unattended":* Ibid.

55 *narratives in Scripture had also been confirmed:* Max E. L. Mallowan, "New Light on Ancient Ur. Excavations at the Site of the City of Abraham Reveal Geographical Evidence of the Biblical Story of the Flood," *National Geographic,* January 1930, 95–130, 44 illus., map.

55 *than on scholarly argument:* B. M. Fagan, *Eyewitness to Discovery,* 131.

55 *the most widely read book on archaeology ever printed:* H. V. F. Winstone, *Woolley of Ur: The Life of Sir Leonard Woolley* (London: Secker and Warburg, 1990), 153.

55 *subsequent trenching at Ur:* G. E. Wright, *Biblical Archaeology* (Philadelphia: Westminster Press, 1957), 119.

55 *such as Tell el Oueli and Choga Mami:* These sites contain the earliest Ubaid culture and reveal a remarkable architecture of post-and-beam–type construction; J. D. Forest, "Tell el-'Oueili Preliminary Report on the 4th Season (1983): Stratigraphy and Architecture," *Sumer* 44, no. 1–2 (1985–86): 55–66; R. J. Mathews and T. J. Wilkinson, "Excavations in Iraq, 1989–90," *Iraq* 53 (1991): 169–82.

55 *failed to encounter this same silt layer:* Max E. L. Mallowan, "Noah's Flood Reconsidered," *Iraq* (1964): 62–82.

55 *In studying nature:* N. Cohn, *Noah's Flood* (New Haven: Yale University Press, 1996), 48.

55 *"the Face of the Earth before the deluge was smooth":* T. Burnet, *Sacred Theory of the Earth* (London: London and Fontwell, 1690; repr. 1965 ed.), 53.

56 *"the great Abysse, the Sea":* Ibid.

56 *during a close encounter with a comet:* W. Whiston's *A New Theory of the Earth* (1755), cited in N. Cohn, *Noah's Flood,* 62–69.

56 *Sunday, October 23, 4004 b.c.:* J. Usher, *The Annals of the Old Testament. From the Beginning of the World* (London: 1658).

56 *"no vestige of a beginning—no prospect for an end":* J. Hutton, *Theory of the Earth.* Proceedings of the Royal Society, 1795.

56 *left no room for the flood:* N. Cohn, *Noah's Flood,* 102.

56 *first edition of* Principles of Geology: C. Lyell, *Principles of Geology: Being an Attempt to Explain the Former Changes in the Earth's Surface by Reference to Causes Now in Operation,* 3 vols. (London: Murray, 1830–33).

56 *"diluvial theory came to be seen":* N. Cohn, *Noah's Flood,* 119.

56 *movement in the United States called Creationism:* G. M. Price, *The Modern Flood Theory of Geology* (New York: Fleming H. Revell, 1935). J. C. Whitcomb and H. M. Morris, *The Genesis Flood: The Biblical Record and Its Scientific Implications* (Grand Rapids, MI: Baker Book House, 1961). R. L. Numbers, *The Creationists* (New York: Alfred A. Knopf, 1992), 319–39.

6. HIDDEN RIVER

Page

62 *Its twenty-first expedition:* R. M. Pratt and F. Workum, "Track Charts, Bathymetry, and Location of Observations, CHAIN Cruise #21: Atlantic Ocean–Mediterranean Sea, August 16, 1961–December 18, 1961" (Woods Hole: Woods Hole Oceanographic Institution, 1962), chart 4 of expedition report.

63 *nearly a mile per second:* In seawater, sound travels at approximately 4,800 feet per second and varies as a function of temperature, pressure, and salinity.

65 *King Darius had led his Persian army:* Herodotus 4.83–89 in D. Grene, *Herodotus: The History* (Chicago: University of Chicago Press, 1987), 311–14.

65 *its excess freshwater:* Ü. Ünlüata et al., "On the Physical Oceanography of the Turkish Straits," in *The Physical Oceanography of Straits,* ed. L. J. Pratt (Deventer, Netherlands: NATO/ASI Series, Kluwer, 1989), 25–29; H. Yüce, "On the Variability of Mediterranean Water Flow into the Black Sea," *Continental Shelf Research* 16, no. 11 (1996): 1399–1413.

65 *The Bosporus Strait and its downstream:* C. G. Gunnerson and E. Oztuvgut, "The Bosphorus," in *The Black Sea: Geology, Chemistry, and Biology,* ed. E. T. Degens and D. A. Ross (Tulsa, OK: Memoir. American Association of Petroleum Geologists, 1974), 119.

65 *exceeds five knots:* R. Scholten, "Role of the Bosphorus in Black Sea

Chemistry and Sedimentation," in *The Black Sea: Geology, Chemistry, and Biology,* ed. E. T. Degens and D. A. Ross (Tulsa, OK: Memoir. American Association of Petroleum Geologists, 1974), 115.

66 *that would drag their boats northward:* cited in C. G. Gunnerson and E. Oztuvgut, "The Bosphorus," 107.

66 *white-painted corks:* L. F. Marsilli, "Observazioni interno al Bosporo traio overo canale di Constantinopoli," cited in Alfred Merz's *Hydrographic Observations in the Bosporus and Dardanelles,* ed. L. Möller, 1928 (Berlin: Proceedings of the Ocean Science Institute of the University of Berlin, 1681), figure 2.

66 *"clashing rocks":* Apollonius of Rhodes, *The Voyage of Argo,* trans. E. V. Rieu, 2nd ed. (London: Penguin Books, 1971), 14.

72 *Borings into these infilling materials:* E. Meriç, ed., *Late Quaternary (Holocene) Bottom Sediments of the Southern Bosphorus and Golden Horn* (Istanbul: Matbaa Teknisyenleri Basımevi Divanyolu, 1990); A. H. Gücüm, "The Bosphorus (Tube-Tunnel) and Golden Horn (Metro) Drillings," in *Late Quaternary (Holocene) Bottom Sediments of the Southern Bosphorus and Golden Horn,* ed. E. Meriç (Istanbul: Matbaa Teknisyenleri Basimevi Divanyolu, 1990), 1–3.

72 *giant blocks of the bedrock, boulders, cobbles, sand, and seashells:* A. S. Derman, "Grain size of the upper sedimentary lithologies (Holocene)" in *Late Quaternary (Holocene) Bottom Sediments of the Southern Bosphorus and Golden Horn,* ed. E. Meriç (Istanbul: Matbaa Teknisyenleri Basimevi Divanyolu, 1990), figure 2.1.

72 *mollusks adapted to salty water:* G. Taner, "The Lamellibranchiata and Gastropods," in *Late Quaternary (Holocene) Bottom Sediments of the Southern Bosphorus and Golden Horn,* ed. E. Meriç (Istanbul: Matbaa Teknisyenleri Basımevi Divanyolu, 1990), 81–85; E. Meriç, M. Sakinç, and O. Eroskay, "Evolution Model of Bosphorus and Goldenhorn Deposits," Mühendislik Jeolojisi Bülteni 10 (1988): 10–14.

7. GIBRALTAR'S WATERFALL
Page

73 *Hit something hard!:* W. B. F. Ryan et al., "Valencia Trough-Site 122," in *Initial Reports of the Deep-Sea Drilling Project,* ed. W. B. F. Ryan and K. J. Hsü (Washington, DC: U.S. Government Printing Office, 1973), 97–98.

75 *Its objective was:* M. N. A. Peterson, "Scientific Goals and Achievements," *Ocean Industry,* (May 1969), 62.

77 *The first sediment core:* W. B. F. Ryan, F. Workum, and J. B. Hersey, "Sediments on the Tyrrhenian Abyssal Plain," *Bulletin, Geological Society of America* 76 (1964): 1275.

79 *only three kinds of rock:* W. B. F. Ryan et al., "Valencia Trough-Site 122," in *Initial Reports of the Deep-Sea Drilling Project,* 103 and figure 10; P. Morrison, *The Ring of Truth: An Inquiry into How We Know What We Know* (New York: Random House, 1987), 162–65.

79 *"turbidity currents":* undersea avalanches of sediments that are carried in suspension and are propelled downslope by gravity. See D. B. Ericson, M. Ewing, and B. C. Heezen, "Deep-sea Sands and Submarine Canyons," *Geological Society of America Bulletin* 62 (1951): 961; B. C. Heezen and M. Ewing, "Turbidity Currents and Submarine Slumps and the 1929 Grand Banks Earthquake," *American Journal of Science* 250 (1952): 849; B. C. Heezen and C. D. Hollister, *The Face of the Deep,* 659 pages (London: Oxford University Press, 1971), 283.

80 *Gessoso Solfifera:* L. Ogniben, "Petrografia della Serie Solfifera Siciliana e considerazioni geologiche relative," *Descriptive Geological Map of Italy Memoir* 33 (1957): 3; L. A. Hardie and H. P. Eugster, "The Depositional Environment of Marine Evaporites: A Case for Shallow, Clastic Accumulation," *Sedimentology* 16 (1971): 188–89.

82 *"the entire Mediterranean might have once dried up":* K. J. Hsü, "When the Mediterranean Dried Up," *Scientific American* (December 1972): 30; K. J. Hsü, *The Mediterranean Was a Desert* (Princeton, NJ: Princeton University Press, 1983), 6.

82 *"We are stymied":* W. B. F. Ryan et al., "Valencia Trough-Site 122," in *Initial Reports of the Deep-Sea Drilling Project,* 104.

83 *"By God, a pillar from Atlantis!":* K. J. Hsü, *The Mediterranean Was a Desert,* 101.

83 *Anhydrite forms now:* D. J. Shearman, "Recent Anhydrite, Gypsum, Dolomite, and Halite from the Coastal Flats of the Arabian Shore of the Persian Gulf," *Proceedings: Geological Society of London* 1607 (1963): 63–65.

83 *called* sabkhas: G. P. Butler, "Modern Evaporite Deposition and Geochemistry of Coexisting Brines, the Sabkha, Trucial Coast, Arabian Gulf," *Journal of Sedimentary Petrology* 39 (1969): 70–89.

84 *"the growth of anhydrite":* W. B. F. Ryan et al., "Strabo Trench and Mountains—Site 129," in *Initial Reports of the Deep-Sea Drilling Project,* ed. W. B. F. Ryan and K. J. Hsü (Washington, DC: U.S. Government Printing Office, 1973), 342.

85 *an abrupt drowning of its desert floor:* K. J. Hsü, W. B. F. Ryan, and M. B. Cita, "Late Miocene Desiccation of the Mediterranean Sea," *Nature* 242 (1973): 242.

86 *rusty-red soil:* W. B. F. Ryan et al., "Boundary of the Sardinia Slope with the Balearic Abyssal Plain," in *Initial Reports of the Deep-Sea Drilling Project,* ed. W. B. F. Ryan and K. J. Hsü (Washington, DC: U.S. Government Printing Office, 1973), 486.

88 *continued another nine hundred feet:* I. S. Chumakov, "Pliocene and Pleistocene Deposits of the Nile Valley in Nubia and Upper Egypt," in *Initial Reports of the Deep-Sea Drilling Project,* ed. W. B. F. Ryan and K. J. Hsü (Washington, DC: U.S. Government Printing Office, 1973), 1242.

90 *transformation from sea to land:* C. Sturani, "A Fossil Eel (*Anguilla* sp.) from the Messinian of Alba (Tertiary Piedmontese Basin). Paleoenvironmental and Paleogeographic Implications," in *Messinian Events in the Mediterranean,* ed. C. W. Drooger (Amsterdam: North-Holland Publishing Company, 1973), 251.

90 *The backbone of this young:* Ibid., 253.

91 *"the evidence for shallow seas":* R. H. Benson, "An Ostracodal View of the Messinian Salinity Crisis," in *Messinian Events in the Mediterranean,* ed. C. W. Drooger (Amsterdam: North-Holland Publishing Company, 1973), 241.

91 *tiny aquatic crustaceans:* R. H. Benson, "Psychrospheric and Continental Ostracodea from Ancient Sediments in the Floor of the Mediterranean," in *Initial Reports of the Deep-Sea Drilling Project,* ed. W. B. F. Ryan and K. J. Hsü (Washington, DC: U.S. Government Printing Office, 1973), 1002–08.

8. VANISHED DESERTS

Page

93 *sea floor spreading:* F. J. Vine, "Spreading of the Ocean Floor—New Evidence," *Science* 154 (1966), 1405; J. R. Heirtzler, "Sea-Floor Spreading," *Scientific American* December 1968): 69.

93 *geological theory:* M. Kay, "North American Geosynclines," *Memoir. Geological Society of America* 48 (1951); J. Aubouin, *Geosynclines* (Amsterdam: Elsevier, 1965).

93 *replaced by the new paradigm:* J. T. Wilson, "Continental Drift," *Scientific American* (April 1963): 86–100.

93 *an attention-gathering paper:* W. C. Pitman and J. R. Heirtzler, "Mag-

netic Anomalies over the Pacific-Antarctic Ridge," *Science* 154, no. 2 (December 1966): 1164.

94 *symmetric anomalies:* W. Sullivan, *Continents in Motion* (New York: McGraw-Hill, 1974), 102; W. Wertenbaker, *The Floor of the Sea: Maurice Ewing and the Search to Understand the Earth* (Boston: Little, Brown and Company, 1974), 197.

94 *the trajectories of the continents:* W. C. Pitman and M. Talwani, "Central North Atlantic Plate Motions," *Science* 174 (1971): 845–48; W. C. Pitman, "Seafloor Spreading in the North Atlantic," *Geological Society of America Bulletin* 83 (1972): 619–43.

95 *Their goal was to document:* J. F. Dewey et al., "Plate Tectonics and the Evolution of the Alpine System," *Geological Society of America Bulletin* 84 (October 1973): 3137–80.

96 *modern humans:* C. B. Stringer, "Documenting the Origin of Modern Humans," in *The Emergence of Modern Humans,* ed. E. Trinkaus (Cambridge: Cambridge University Press, 1989): 67–96.

96 *apparently emerged from Africa:* A. C. Wilson and R. L. Cann, "The Recent African Genesis of Humans," *Scientific American,* (April 1992): 68–73.

96 *"were rare, but real events":* K. J. Hsü, M. B. Cita, and W. B. F. Ryan, eds., "The Origin of the Mediterranean Evaporite," vol. 13, *Initial Reports of the Deep-Sea Drilling Project* (Washington, DC: U.S. Government Printing Office, 1973), 1228.

97 *Chumakov's hidden Nile gorge:* V. Chumakov, "Pliocene and Pleistocene Deposits of the Nile Valley in Nubia and Upper Egypt," in *Initial Reports of the Deep-Sea Drilling Project,* ed. W. B. F. Ryan and K. J. Hsü (Washington, DC: U.S. Government Printing Office, 1973), 1242–43.

97 *Beneath Cairo, the ancestral Nile:* W. B. F. Ryan and M. B. Cita, "Messinian Badlands on the Southeastern Margin of the Mediterranean Sea," *Marine Geology* 27 (1978): 349–63.

99 *Only one, Porites:* M. Estaban, "Significance of Upper Miocene Coral Reefs of the Western Mediterranean," *Paleogeography, Paleoclimatology, Paleoecology* 29 (1980): 169–88.

99 *such as Sicily, Sardinia, and Malta:* A. Azzaroli, "Cainozoic Mammals and the Biogeography of the Island of Sardinia, Western Mediterranean," *Paleogeography, Paleoclimatology, Paleoecology* 36 (1981): 107–11; A. Azzaroli and G. Guazzone, "Terrestrial Mammals and Land Connections in the Mediterranean Before and During the Messinian," *Palaeogeography, Palaeoclimatology, Palaeoecology* 29 (1980): 155–67.

9. PONTUS AXENUS

Page

101 *"a land-locked basin":* J. N. Wilford, "Drillings Hint Mediterranean Gets Smaller," *The New York Times,* Saturday, Oct. 10, 1970.

101 *recent sedimentary history of the Black Sea:* D. A. Ross, E. T. Degens, and J. MacIlvaine, "Black Sea: Recent Sedimentary History," *Science* 170, no. 9 (Oct. 1970): 163–65.

101 *the freshwater species had disappeared:* N. G. Mayard, "Diatoms in Pleistocene Deep Black Sea Sediments," in *The Black Sea—Geology, Chemistry, and Biology,* ed. E. T. Degens and D.A. Ross (Tulsa, OK: Memoir. American Association of Petroleum Geologists, 1974), 389–95; D. Wall and B. Dale, "Dinoflagellates in the Late Quaternary Deep-Water Sediments of the Black Sea," in *The Black Sea—Geology, Chemistry, and Biology,* ed. E. T. Degens and D. A. Ross (Tulsa, OK: Memoir. American Association of Petroleum Geologists, 1974), 364–80.

102 *for thousands of years:* E. T. Degens and D. A. Ross, "Chronology of the Black Sea over the Last 25,000 Years," *Chemical Geology* 10 (1972): 1–16.

102 *reconnected with the Aegean arm:* D. J. Stanley and C. Blanpied, "Late Quaternary Exchange Between the Eastern Mediterranean and the Black Sea," *Nature* 285, no. 19 (June 1980): 537–41.

102 *the mid-eighth century B.C.:* G. A. Jones, "Tales of Black Sea Sedimentation, Exploration and Colonization," *AMS Pulse* 2, (Summer 1993): 1–6; G. A. Jones, "A New Hypothesis for the Holocene Appearance of Coccolithophores in the Black Sea," *AMS Pulse* supplement (1993): 195, see figure 1.

103 *"having swallowed up a large area of land which retreated before it":* Pliny the Elder, *Natural History,* trans. J. F. Healy (London: Penguin Books, 1991): 51.

103 *He gave it the name Pontus Axenus:* Ibid., sentence 1 of Book 5 on p. 63.

103 *some earlier discoveries:* D. A. Ross, "The Red and the Black Seas," *American Scientist* 59, (July/August 1971): 420–24.

103 *they would go left into the Aegean and on to the Black Sea:* interview with David Ross at Woods Hole, Massachusetts, January, 1995.

104 *a dark black, jellylike mud:* D. A. Ross and E. T. Degens, "Recent Sediments of the Black Sea," in *The Black Sea—Geology, Chemistry, and Biology,* ed. E. T. Degens and D. A. Ross (Tulsa, OK: Memoir. American Association of Petroleum Geologists, 1974), 183–99.

104 *revealed hair-thin white bands:* E. T. Degens and D. A. Ross, "Oceano-

graphic Expedition in the Black Sea," *Naturwissenschaften* 57 (1970): 351, figure 3a.

104 *the skeletal framework:* D. Bukry et al., "Geological Significance of Coccoliths in Fine-Grained Carbonate Bands of Postglacial Black Sea Sediments," *Nature* 226, no. 11 (April 1970): 157, figure 2.

104 *membranes of proteinlike structures:* E. T. Degens, S. W. Watons, and C. C. Remsen, "Fossil Membranes in Cell Wall Fragments from a 7,000-Year-Old Black Sea Sediment," *Science* 168 (1970): 1207–08.

105 *snails diagnostic of a terrestrial streambed:* G. I. Popov, "New Data on the Stratigraphy of Quaternary Marine Sediments of the Kerch Strait," *Transactions of the USSR Academy of Sciences* 213, no. 4 (1973): 84–86.

106 *overlying a former terrestrial landscape:* P. N. Kuprin, F. A. Scherbakov, and I. I. Morgunov, "Correlation, Age, and Distribution of the Postglacial Continental Terrace Sediments of the Black Sea," *Baltica* 5 (1974): 241–49.

106 *beach deposits:* F. A. Shcherbakov et al., *Sedimentation on the Continental Shelf of the Black Sea* (Moscow: Nauka Press, 1978).

106 *Rivers flowing into the falling surface of this body of water had deepened:* A. B. Ostrovskiy et al., "New Data on the Paleohydrological Regime of the Black Sea in the Upper Pleistocene and Holocene," in *Paleogeography and Deposits of the Pleistocene of the Southern Seas of the USSR,* ed. P. A. Kaplin and F. A. Shcherbakov (Moscow: Nauka Press, 1977), 131–41.

106 *the replacement marine fauna:* L. A. Nevesskaya, "Late Quaternary Bivalve Mollusks of the Black Sea: Their Systematics and Ecology," *Transactions of the Institute of Paleontology of the USSR Academy of Sciences* 105 (1965): 1–390; K. M. Shimkus, V. V. Mukhina, and E. S. Trimonis, "On the Role of Diatoms in Late Quaternary Sedimentation in the Black Sea," *Oceanology, USSR Academy of Sciences* 13 (1973): 1066–71.

106 *submergence of the Black Sea margin:* P. V. Federov, "Postglacial Transgression of the Black Sea," *International Geology Review* 14, no. 2 (1971): 160–64; Y. N. Nevesskiy, "The Question of the Recent Transgression of the Black Sea," 26 (1958); Y. N. Nevesskiy, "Postglacial Transgression of the Black Sea," *Transactions of the Institute of Paleontology of the USSR Academy of Sciences* 28 (1961): 317–20.

106 *below its outlet:* A. L. Chepalyga, "Inland Sea Basins," in *Late Quaternary Environments of the Soviet Union,* ed. H. E. Wright, Jr., and C. W. Barnosky (Minneapolis: University of Minnesota Press, 1984), 229–47.

107 *the past climate of the Black Sea periphery:* D. D. Kvasov, "Paleohydrol-

ogy of Eastern Europe in Late Quaternary Time," in *Early Publication of a Lecture in Memory of L. S. Berga* (Leningrad: Izdatelstvo Nauka, 1968), 65–81; D. D. Kvasov, *Late Quaternary History of Major Lakes and Inland Seas of Eastern Europe* (Leningrad: Nauka Press, 1975); M. B. Chernyshova, "Palynological Studies of Bottom Sediments from the Continental Terrace," in *Geological-Geophysical Studies of the Bulgarian Sector of the Black Sea,* ed. P. N. Kuprin (Sofia: Bulgarian Academy of Sciences, 1980), 213–22; A. Traverse, "Palynologic Investigation of Two Black Sea Cores," in *The Black Sea—Geology, Chemistry, and Biology,* ed. E. T. Degens and D. A. Ross (Tulsa, OK: American Association of Petroleum Geologists, 1974), 381–88.

10. RED HILL

Page

109 *a comprehensive synthesis:* V. Sibrava, *Survey of Czechoslovak Quaternary,* vol. 34 (Warsaw: Archaeological Institute of the Czechoslovak Academy of Sciences, 1961).

109 *Evolution of Shorelines:* R. W. Fairbridge, "Quaternary Shoreline Problems at Inqua, 1969," *Quaternaria* 15 (1971): 1–18.

110 *"an agent capable of producing a change of 350 feet":* C. Maclaren, "The Glacial Theory of Prof. Agassiz of Neuchâtel," *American Journal of Science* 42 (1842): 346–65.

110 *by the Institute for Nuclear Studies:* W. F. Libby, *Radiocarbon Dating* (Chicago: University of Chicago Press, 1952).

110 *shifting shorelines could be dated:* W. S. Broecker and E. A. Olson, "Lamont Radiocarbon Measurements IV," *American Journal of Science* 1, radiocarbon supplement (1959): 111–32.

110 *Carbon 14 decayed:* Radiocarbon dating is a means of obtaining the age of organic materials such as bone, charcoal, wood, peat, or shell from an organism that lived in the past. Carbon is most commonly present on the Earth as an atom with six protons and six neutrons in its nucleus, giving this chemical species an atomic weight of 12. However, there is a relatively rare variant (known as an isotope) of carbon with six protons and eight neutrons in its nucleus, making an atomic weight of 14. Such carbon 14 (or C14) is produced naturally in the upper atmosphere as the result of cosmic ray bombardment. Newly generated carbon 14 becomes rapidly distributed throughout the atmosphere. It mixes over the span of hundreds of years through the ocean, whereas a small percentage of the Earth's carbon 14 reservoir immediately enters the terrestrial biosphere. Once a living organism dies, metabolic pro-

cesses cease, and it no longer takes up new carbon 14. The fossil tissue or skeleton acts like a sealed tomb in which the unstable carbon 14 gradually disappears by radioactive decay processes while the stable carbon 12 remains. The rate of the decay, referenced as a half-life, is such that only 50 percent of the original amount of the radioactive isotope of carbon is left 5,730 years after the death of the organism. A radiocarbon date is obtained by measuring the amount of remaining carbon 14 in proportion to the nondecaying carbon 12. A radiocarbon age is conventionally expressed in years before present (years BP) or, specifically, in years before A.D. 1950, the zero-age reference adopted after Willard Libby's first successful exploitation of the method in 1949. The production of new carbon 14 in the upper atmosphere is not a steady process; therefore, conventional carbon 14 years do not align directly with calendar years, and calibrations need to be applied.

110 *reversals of the earth's magnetic field:* The earth's magnetic field reverses its polarity at a nonperiodic frequency averaging about a quarter of a million years. See: A. Cox, "Geomagnetic Reversals," *Science* 163 (1969): 237–45; W. Glenn, *The Road to Jaramillo* (Stanford: Stanford University Press, 1982), 93–117.

111 *for an astronomical theory of the Ice Age:* M. M. Milankovitch, "Canon of Insolation and the Ice Age Problem," *Royal Serbian Academy, Belgrade* 132 (1941); M. M. Milankovitch, "Astronomical Theory of the Ice Ages," in *Handbook of Geophysics,* ed. W. Koppen and G. Geiger (Berlin: 1938), 593.

111 *"Why don't you tag along?":* Interview with Jirí Kukla in Blauvelt, New York, June, 1994. The others present at the dinner in Paris included John Imbrie and James Hays. In 1976 they would publish a seminal paper demonstrating the validity of the Milankovitch theory to explain the entire Quaternary ice age (see J. D. Hays, J. Imbrie, and N. J. Shackleton, "Variations in the Earth's Orbit: Pacemaker of the Ice Ages," *Science* 194, no. 10 (December 1976): 1121–32.

112 *the new findings:* W. S. Broecker et al., "Milankovitch Hypothesis Supported by Precise Dating of Coral Reefs and Deep-Sea Sediments," *Science* 159, no. 19 (January 1968): 297–300; W. S. Broecker and J. van Donk, "Insolation Changes, Ice Volumes and the O^{18} Record in Deep-Sea Cores," *Reviews of Geophysics and Space Physics* 8 (1970): 169–97.

112 *one sediment core:* D. B. Ericson and G. Wollin, *The Deep and the Past* (New York: Alfred A. Knopf, 1964); D. B. Ericson and G. Wollin, *The Ever-Changing Sea* (New York: Alfred A. Knopf, 1967), 173–202.

112 *the former estate of Thomas Lamont:* W. Wertenbaker, *The Floor of the Sea: Maurice Ewing and the Search to Understand the Earth* (Boston: Little, Brown and Company, 1974).

112 *the Lamont Geological Observatory:* later in 1969 to become the Lamont-Doherty Geological Observatory and then in 1993 the Lamont-Doherty Earth Observatory.

112 *magnetic reversals in the deep-sea cores:* N. D. Opdyke et al., "Paleomagnetic Study of Antarctic Deep-Sea Cores," *Science* 154, no. 4 (November 1966): 349–57; B. P. Glass et al., "Geomagnetic Reversals and Pleistocene Chronology," *Nature* 216, no. 5114 (1967): 437–42.

113 *recently bound book:* J. Demek and Jirí Kukla, *The Periglacial Zone, Loess and Paleosoils of Czechoslovakia* (Brno: Czechoslovak Academy of Sciences, 1969), figure 32 on p. 93.

113 *using magnetic reversals for dating Ice Age climates:* V. Bucha et al., "Paläomagnetische Messungen in Lössen," in *The Periglacial Zone, Loess and Paleosoils of Czechoslovakia,* ed. J. Demek and J. Kukla (Brno: Tschechoslowakische Akademie der Wissenschaften, 1969), 123–32.

113 *Dolni Vestonice:* M. Gimbutas, *The Language of the Goddess,* (New York: HarperCollins, 1991), 31, 51; R. J. Wenke, *Patterns in Prehistory,* 3rd ed. (Oxford: Oxford University Press, 1990), 178, figure 4.17.

113 *sketches of the campsites:* J. Demek and Jirí Kukla, *The Periglacial Zone, Loess and Paleosoils of Czechoslovakia,* figure 42, p. 115.

113 *"Venus figurines":* three-dimensional carvings in rock, ivory, bone, and fired clay as well as two-dimensional incisions and paintings on cave walls with realistic renditions of the female body, generally without facial detail but portraying large, voluptuous breasts, a protruding (interpreted as pregnant) belly, and often in association with the image of childbirth. "Venus's sculptures are thought to be symbols of fertility and magic. Some specimens from Ukraine have body decoration suggestive of clothing and tattoos. See P. Rice, "Prehistoric Venuses: Symbols of Motherhood," *Journal of Anthropological Archaeology* 37 (1981): 402–14; M. Gimbutas, *The Gods and Goddesses of Old Europe* (London: Thames and Hudson, 1982).

114 *giant sand dunes sculpted by wind:* J. Sekyra, "Wind-blown Sands," in *Survey of Czechoslovak Quaternary,* ed. V. Sibrava (Warsaw: 1961), 29–38.

115 *a set of exposed reefs:* W. S. Broecker et al., "Milankovitch Hypothesis

Supported by Precise Dating of Coral Reefs and Deep-Sea Sediments," *Science* 159, no. 19 (January 1968): 297–300.

115 *minute quantities of uranium:* W. S. Broecker and D. L. Thurber, "Uranium-Series Dating of Corals and Oolites from Bahamas and Florida Key Limestones," *Science* 149, no. 3679 (1965): 58–60.

116 *nine thousand years before the present:* E. T. Degens and D. A. Ross, "Chronology of the Black Sea over the last 25,000 Years," *Chemical Geology* 10 (1972): 1–16.

11. AQUANAUTS

Page

118 *Dimitrov reported on a series of sediment cores:* P. S. Dimitrov, "Radiocarbon Datings of Bottom Sediments from the Bulgarian Black Sea Shelf," *Oceanology, Bulgarian Academy of Sciences* 9 (1982): 45–53; P. S. Dimitrov, L. I. Govberg, and V. Il. Kuneva-Abadzhieva, "Marine Quaternary Deposits of the Peripheral region of the Shelf from the Western Part of the Black Sea," *Oceanology, Bulgarian Academy of Sciences* 5 (1979): 67–78.

120 *the catastrophic explosion of the Chernobyl nuclear power plant:* A. I. Nikitin, V. I. Medinets, and V. V. Chumichev, "Radioactive Pollution in the Black Sea in October, 1986, Caused by the Chernobyl Accident," *Atomic Energy* 65, no. 2 (1986): 134–37; M. Goldman, "Chernobyl: A Radiobiological Perspective," *Science* 238 (1987): 622–23.

120 *wreck site of the* Titanic: W. B. F. Ryan, "The use of mid-range side-looking sonar to locate the wreck of the *Titanic,"* *Subtech* 83 (London: 1983), Paper No. 11.4, 1–14.

124 *drowned path of the ancient Don River:* G. I. Popov, "New Data on the Stratigraphy of Quaternary Marine Sediments of the Kerch Strait," *Transactions of the USSR Academy of Sciences* 213, no. 4 (1973): 84–86; S. I. Skiba, F. A. Shcherbakov, and P. N. Kuprin, "On Paleogeography of the Kerch-Taman Region in Late Pleistocene and Holocene," *Oceanologia* 15, no. 5 (1975): 862–67; A. B. Ostrovskiy et al., "New data on the Stratigraphy and Geochronology of Pleistocene Marine Terraces of the Black Sea Coast, Caucasus, and Kerch-Taman Region," in *Paleogeography and Deposits of the Pleistocene of the Southern Seas of the USSR,* ed. P. A. Kaplin and F. A. Shcherbakov (Moscow: Nauka Press, 1977), 61–69.

126 *discharging a plume of bubbles:* V. N. Moskalenko, W. C. Pitman, and W. B. F. Ryan, "Discerning of Gas Saturated Sediments on Seismoacous-

tic Sections of the Northwestern Black Sea," *Oceanology* 36, no. 3 (1996): 462–69.

12. IMMIGRANTS

Page

128 *radioactive nucleides released from the Chernobyl accident:* K. O. Buesseler et al., "Scavenging and Particle Deposition in the Southwestern Black Sea—Evidence from Chernobyl Radiotracers," *Deep-Sea Research* 37 (1990): 413–30; E. A. Kontar et al., "Fate of Chernobyl Radionuclides in Black Sea sediments" (in press).

129 *This first core:* latitude 44°57.6′ N, longitude 32°05.5′ E, depth 68 meters, 134 centimeters in length.

132 *At the eighth coring location:* latitude 44°54.4′ N, longitude 32°08.5′ E, depth 99 meters, 159 centimeters in length.

133 *She found cracks:* C. O. Major, "Late Quaternary Sedimentation in the Kerch Area of the Black Sea Shelf: Response to Sea Level Fluctuation" (BA thesis, Wesleyan University, 1994), figure 3.12, p. 71.

13. CLOSE ENCOUNTER

Page

136 *In 1983, Bill Ryan had been among the first civilian oceanographers:* The early deployment of the Global Positioning System was aboard the Research Vessel *Thomas Washington* of the Scripps Institution of Oceanography, University of California, surveying along the East Pacific Rise on expedition PASCOIWT and with funding from the Office of Naval Research.

137 *The survey expanded:* see W. Ryan et al., *Cruise Report—Black Sea Shelf: Research Vessel Aquanaut, June 10–22, 1993* (Gelendzhik: Southern Branch of P. P. Shirshov Institute of Oceanology, 1993).

14. BEACHCOMBERS

Page

143 *a manuscript he had submitted to the British journal* Deep-Sea Research: G. A. Jones and A. R. Gagnon, "Radiocarbon Chronology of Black Sea Sediments," *Deep-Sea Research* 41, no. 3 (1994): 531–57.

143 *alternating dark and light layers of sediment:* E. T. Degens et al., "Varve Chronology: Estimated Rates of Sedimentation in the Black Sea Deep Basin," in *Initial Reports of the Deep-Sea Drilling Project,* ed. D. A. Ross (Washington, DC: U.S. Government Printing Office, 1980).

143 *many other adherents as well:* B. J. Hay et al., "Sediment Deposition in

the Late Holocene Abyssal Black Sea with Climatic and Chronological Implications," *Deep-Sea Research* 38, supplement 2 (1991): S1211–35; M. A. Arthur et al., "Varve Calibrated Records of Carbonate and Organic Carbon Accumulation over the Last 2000 Years in the Black Sea," *Global Biogeochemical Cycles* 8 (1994): 195–217.

144 *Its precision had steadily improved:* G. A. Jones et al., "High-Precision AMS Radiocarbon Measurements of Central Arctic Ocean Seawaters," *Nuclear Instruments and Methods* B92 (1994): 426–30.

144 *specimen of acacia wood:* V. Kaharl, "A Reexamination of the World's First 14C Analysis," *The National Ocean Science Accelerator Mass Spectrometry Facility Newsletter* 1 (Fall 1992): 1–6.

145 *The detonations of Krakatoa:* M. N. Padang, "Two Catastrophic Eruptions in Indonesia, Comparable with the Plinian Outburst of the Volcano of Thera (Santorini) in Minoan Time," in *Acta of the 1st International Scientific Congress on the Volcano of Thera,* ed. A. Kaloyeropoyloy (Athens: Archaeological Services of Greece, 1971), 51–63.

145 *Santorini jetted a plume of ash:* D. Ninkovich and B. C. Heezen, "Santorini Tephra," paper presented at the Submarine Geology and Geophysics: Proceedings of the Seventeenth Symposium of the Colston Research Society (London, 1965), 413–53.

145 *to the Nile Delta:* D. J. Stanley and H. Sheng, "Volcanic Shards from Santorini (Upper Minoan Ash) in the Nile Delta, Egypt," *Nature* 320 (1986): 733–35.

145 *northeast to blanket the floor of the Black Sea:* D. G. Sullivan, "The Discovery of Santorini Minoan Tephra in Western Turkey," *Nature* 333 (1988): 552–54; F. Guichard et al., "Tephra from the Minoan Eruption of Santorini in Sediments of the Black Sea," *Nature* 363, no. 17 (June 1993): 610–12.

145 *Stunted fossil tree rings:* V. V. LaMarche, Jr., and K. K. Hirschboeck, "Frost Rings in Trees as Records of Major Volcanic Eruptions," *Nature* 307 (1984): 121–26; M. G. L. Baille and M. Munro, "Irish Tree Rings, Santorini and Volcanic Dust Veils," *Nature* 331 (1988): 344–46.

145 *AMS measurements on seeds of grains:* W. L. Friedrich, P. Wagner, and H. Tauber, "Radiocarbon Dated Plant Remains from Akrotiri Excavation on Santorini, Greece," in *Thera and the Aegean World III,* ed. D. A. Hardy and A. C. Renfrew, Proceedings of the Third International Congress, Santorini, Greece (London: The Thera Foundation, 1990), 188–96; P. I. Kuniholm, "Overview and Assessment of the Evidence for the Date of the Eruption of Thera," in *Thera and the Aegean World III,* ed. D. A. Hardy and A. C. Renfrew, Proceedings of the Third International

Congress, Santorini, Greece (London: The Thera Foundation, 1990), 13–18.

145 *Minoan village of Akrotiri:* S. Marinatos, *Excavations at Thera I-V* (Athens: Athens Archaeological Society, 1968–1972).

147 *by the tiny aquatic algae named* Emiliania huxleyi: D. Bukry, "Coccoliths as Paleosalinity Indicators—Evidence from the Black Sea," in *The Black Sea—Geology, Chemistry, and Biology,* ed. E. T. Degens and D. A. Ross (Tulsa, OK: Memoir. American Association of Petroleum Geologists, 1974), 333–63; D. Bukry et al., "Geological Significance of Coccoliths in Fine-Grained Carbonate Bands of Past Glacial Black Sea Sediments," *Nature* 226, no. 11 (April 1970): 156–58.

147 *first commercial foothold along the Black Sea shore:* G. A. Jones, "Tales of Black Sea Sedimentation, Exploration, and Colonization," *AMS Pulse* 2, (Summer 1993): 1–6.

147 *trip as a stowaway in the bilgewater of Greek rowing ships:* G. A. Jones, "A New Hypothesis for the Holocene Appearance of Coccolithophores in the Black Sea" (abstract), *AMS Pulse, supplement* (1993), 5.

148 *into the Great Lakes of central North America:* D. V. Cohn, "Foreign Invaders Arrive in Bay by Ship, Regional Officials Told," *The Washington Post,* January 7 1994, p. 3.

148 *Conventional Age, Reservoir Age, and Calibrated Age:* Jones presented the dates he had obtained three ways. The first was raw carbon 14 years determined directly from his measurement of the ratio of carbon 12 and carbon 14 atoms remaining in the seashells. Second, he made a correction for what is called the "reservoir effect." Whereas carbon 14 atoms newly produced in the stratosphere from cosmic ray collisions rapidly mix throughout the global atmosphere in less than a decade, they enter the ocean more slowly. Thus, seawater has an age some hundreds of years older than the atmosphere. Marine organisms that consume seawater bicarbonate to make shell material do this long after the radioactive decay in seawater has begun. Jones determined this aging effect by dating the difference between terrestrial (extracting carbon from the air) and marine organisms (extracting carbon from seawater) using seashells collected from living specimens that he obtained from Russian museums. He then compared his measured dates on these shells with the dates of their collection and obtained a difference of 480 years. He checked this difference, and it agreed with the age discrepancy he got by dating the volcanic ash derived from the late Bronze Age eruption of the island of Santorini in the nearby Aegean and independent carbon 14 dates from individual plant seeds, grains, and nuts

entombed in storage jars beneath the ash fall. In the third and final step he calibrated his reservoir corrected dates to calendar ages using the modern technique of dendrochronology. Jones obtained this latter calibration from other researchers who had radiocarbon-dated individual tree rings of bristlecone pine trees going back thousands of years and had compared those ages with actual tree ring counts. The dendrochronological calibration adjusts for an uneven production of carbon 14 in the atmosphere through the millennia. Each step of correction and calibration introduces new uncertainties and has the collective effect of widening the error limits of the final calendar age. See M. Stuiver and T. F. Braziuna, "Modelling Atmospheric [14]C Influences and [14]C Ages of Marine Samples to 10,000 B.C.," *Radiocarbon* 35 (1993): 137–89.

149 *A geochemist at his institution had calculated:* W. G. Deuser, "Evolution of Anoxic Conditions in the Black Sea During the Holocene," in *The Black Sea—Geology, Chemistry, and Biology,* ed. E. T. Degens and D. A. Ross (Tulsa, OK: Memoir. American Association of Petroleum Geologists, 1974), 133–36; W. G. Deuser, "Late-Pleistocene and Holocene History of the Black Sea as Indicated by Stable Isotope Studies," *Journal of Geophysical Research* 77 (1972): 1071–77.

150 *that farming had been practiced:* G. Barker, *Prehistoric Farming in Europe* (Cambridge: Cambridge University Press, 1985); I. Hodder, *The Domestication of Europe* (Oxford: Blackwell Publishers, 1990).

15. BACK OF THE ENVELOPE

Page

153 *"isostatic subsidence":* W. R. Peltier, "Ice Age Paleotopography," *Science* 265, no. 8 (July 1994): 195–201.

154 *away from the area where the load is applied:* P. Kearey and F. J. Vine, *Global Tectonics* (Oxford: Blackwell Scientific Publications, 1990): 34–36, figure 2.32.

154 *two experts from the oil patch:* Bill Svendsen of Longyear Co. and Ken Taylor of Fugro McClelland Inc.

154 *prepared Cajun style:* interview with Richard Fairbanks in 1994.

155 *nearly twenty thousand years ago:* R. G. Fairbanks, "A 17,000-Year Glacio-Eustatic Sea Level Record: Influence of Glacial Melting Rates on the Younger Dryas Event and Deep-Ocean Circulation," *Nature* 342, no. 7 (December 1989): 637–42.

155 *common Elk Horn variety known as* Acropora palmata: R. G. Lighty, I. G. Macintyre, and R. Stuckenrath, *"Acropora Palmata* Reef Frame-

work: A Reliable Indicator of Sea Level in the Western Atlantic for the Past 10,000 Years," *Coral Reefs* 1 (1982): 125–30.

155 *to carry out the mass spectrometry:* E. Bard et al., "U/Th and 14C Ages of Corals from Barbados and Their Use for Calibrating the 14C Time Scale Beyond 9000 Years B.P.," *Nuclear Instruments and Methods in Physics Research* B52 (1990): 461–68.

155 *"an agent capable of producing a change of three hundred and fifty feet on the level of the sea":* C. Maclaren, "The Glacial Theory of Prof. Agassiz of Neuchâtel," *American Journal of Science* 42 (1842): 346–65.

156 *The first of the rapid pulses:* R. G. Fairbanks, "The Age and Origin of the 'Younger Dryas Climate Event' in Greenland Ice Cores," *Paleoceanography* 5, no. 6 (1990): figure 3, p. 943.

156 *the Upper Dnieper, Upper Volga, Dvina-Pechora, Tungusta, Pur, and Mansi:* M. G. Grosswald, "Late Weichselian Ice Sheet of Northern Eurasia," *Quaternary Research* 13 (1980): figure 7, p. 16.

157 *"New Euxine" deposits:* A. L. Chepalyga, "Inland Sea Basins," in *Late Quaternary Environments of the Soviet Union,* ed. H. E. Wright, Jr., and C. W. Barnosky (Minneapolis: University of Minnesota Press, 1984), 229–47.

157 *The second meltwater spike:* R. G. Fairbanks, "The Age and Origin of the 'Younger Dryas Climate Event' " figure 3, p. 943.

157 *away from the Black Sea and westward across Poland:* M. G. Grosswald, "Late Weichselian Ice Sheet," 28.

159 *microfossils from new boreholes:* E. Meriç and M. Sakinç, "Foraminifera," in *Late Quaternary (Holocene) Bottom Sediments of the Southern Bosphorus and Golden Horn,* ed. E. Meriç (Istanbul: Matbaa Teknisyenleri Basımevi Divanyolu, 1990), 13–26.

159 *well known internationally:* H. Yüce, "Investigation of the Mediterranean Water in the Strait of Istanbul (Bosphorus) and the Black Sea," *Oceanologica Acta* 13, no. 2 (1990): 177–86; H. Yüce, "On the Variability of Mediterranean Water Flow into the Black Sea," *Continental Shelf Research* 16, no. 11 (1996): 1399–1413.

159 *A simple formula:* $Q = AR^{2/3} S^{1/2} n^{-1}$ where Q = the discharge rate, A = the channel cross-sectional area, R = the hydraulic radius, S = the slope, and n = a roughness coefficient used for friction.

16. ANYBODY THERE?

Page

165 *For over seventy years:* R. Pumpelly, *Explorations in Turkestan. Expedition of 1904. Prehistoric Civilizations of Anau; Origins, Growth, and*

Influence of Environment, 2 vols. (Washington, DC: Carnegie Institution of Washington, 1908).

165 *"conquer new means of support"*: Ibid., 65–66.

165 *"oasis theory of agricultural origins"*: P. J. Watson, "Origins of Food Production in Western Asia and Eastern North America: A Consideration of Interdisciplinary Research in Anthropology and Archaeology," in *Quaternary Landscapes,* ed. L. C. K. Shane and E. J. Cushing (Minneapolis: University of Minnesota Press, 1991), 4.

165 *in succeeding decades it caught the attention:* B. G. Trigger, *Gordon Childe: Revolutions in Archaeology* (London: Thames and Hudson, 1980), 29.

166 *"the greatest prehistorian in Britain"*: S. Piggott, "Vere Gordon Childe, 1892–1957," *Proceedings of the British Academy* 44 (1958): 312.

167 *"a revolution whereby man ceased to be purely a parasite"*: V. G. Childe, *The Most Ancient East: the Oriental Prelude to European Prehistory* (London: Kegan, Paul, Trench & Trubner, 1928); quoted in B. M. Fagan, *Eyewitness to Discovery* (Oxford: Oxford University Press, 1996), 96.

167 *the oldest domesticated sheep in Europe:* V. G. Childe, *New Light on the Most Ancient East,* 4th ed. (London: Routledge and Kegan Paul, 1952), 26.

167 *he regarded his field of archaeology:* V. G. Childe, *Piecing Together the Past: The Interpretation of Archaeological Data* (London: Routledge and Kegan Paul, 1956); B. G. Trigger, *Gordon Childe: Revolutions in Archaeology* (London: Thames and Hudson, 1980), 134.

168 *He extended his thesis:* V. G. Childe, *Society and Knowledge: The Growth of Human Traditions* (New York: Harper, 1956), 94.

168 *As a result, his reconstruction of Europe's past was seriously challenged:* B. G. Trigger, *Gordon Childe: Revolutions in Archaeology,* 168.

168 *along the banks of rivers:* A. Sherratt, "The Development of Neolithic and Copper Age Settlement in the Great Hungarian Plain. Part 2: Site Survey and Settlement Dynamics," *Oxford Journal of Archaeology* 2 (1983): 12–41; A. Sherratt, "The Development of Neolithic and Copper Age Settlement in the Great Hungarian Plain. Part 1: The Regional Setting," *Oxford Journal of Archaeology* 1 (1982): 287–316.

170 *the site had been used for hundreds of years:* five levels of occupation are present in Lepenski Vir phases one and two prior to any signs of farming.

170 *Dragoslav Srejovic, the discoverer and first excavator of Lepenski Vir:* D. Srejovic, "Europe's First Monumental Sculpture: New Discoveries at

Lepenski Vir," in *New Aspects of Antiquity,* ed. M. Wheeler (London: Thames and Hudson, 1972).

170 *aspiration to control death and tame the wild in nature:* I. Hodder, *The Domestication of Europe* (Oxford: Blackwell, 1990), 25.

171 *Harris had just completed the editing:* D. R. Harris, "Introduction: Themes and Concepts in the Study of Early Agriculture," in *The Origins and Spread of Agriculture and Pastoralism in Eurasia,* ed. D. R. Harris (London: UCL Press, 1996), pp. 1–9; D. R. Harris, "The Origins and Spread of Agriculture and Pastoralism in Eurasia: An Overview," in *The Origins and Spread of Agriculture and Pastoralism in Eurasia,* ed. D. R. Harris (London: UCL Press, 1996), 552–73.

173 *"the homes of primeval agriculture":* N. I. Vavilov, *Studies on the Origins of Cultivated Plants* (Leningrad: Institut Botanique Appliqué et d'Amélioration des Plants, 1926), 219.

171 *"agriculture originated independently only very rarely":* D. R. Harris, "The Origins and Spread of Agriculture and Pastoralism in Eurasia: An Overview," 554.

171 *"rift-valley oases":* Ibid., 554.

171 *out of the Taurus and Zagros mountains of Anatolia and Persia:* F. Hole, "The Context of Caprine Domestication in the Zagros Region," in *The Origins and Spread of Agriculture and Pastoralism in Eurasia,* ed. D. R. Harris (London: UCL Press, 1996), 263–81.

171 *environmental change—particularly climate—was instrumental:* Ibid., 555.

172 *shift back to near-glacial conditions:* C. Vergnasud-Grazzini, W. B. F. Ryan, and M. B. Cita, "Stable isotope fractionation, climate change and episodic stagnation in the eastern Mediterranean during the late Quaternary," *Marine Micropaleontology* 2 (1977): figure 2, 356.

172 *Called the Younger Dryas:* J. Mangerud et al., "Quaternary Stratigraphy of Norden, a Proposal for Terminology and Classification," *Boreas* 3 (1974): 109–28.

172 *it lasted from 10,500 to 9,400 B.C.:* S. Björck et al., "Synchronized Terrestrial Atmospheric Deglacial Records Around the Atlantic," *Science* 274 (November 1996): 1155–60.

172 *One of the first archaeologists who recognized:* The account is presented by its principal players in A. M. T. Moore, G. C. Hillman, and A. J. Legge, *Abu Hureyra and the Advent of Agriculture* (Oxford: Oxford University Press, in press).

172 *two successive villages stacked one upon the other:* A. M. T. Moore and

G. C. Hillman, "The Excavation of Tell Abu Hureyra in Syria: A Preliminary Report," *Proceedings of the Prehistoric Society* 41 (1975): 50–77.

173 *wild barley, lentil, and vetch, and the fruit of the hackberry:* Found in Appendix A: "The Plant Remains of Tell Abu Hureyra: A Preliminary Report" by G. Hillman in A. M. T. Moore and G. C. Hillman, "The Excavation of Tell Abu Hureyra in Syria: A Preliminary Report," *Proceedings of the Prehistoric Society* 41 (1975): 70–73.

173 *In the deteriorating climate of the Younger Dryas:* A. M. T. Moore and G. C. Hillman, "The Pleistocene to Holocene Transition and Human Economy in Southwest Asia: The Impact of the Younger Dryas," *American Antiquity* 57, no. 3 (1992): 482–94.

174 *The artifacts and bone residues conveyed the message:* M. Magaritz, G. A. Goodfriend, in *Abrupt Climatic Change* eds. W. H. Berger and L. D. Labeyrie (Boston: D. Reidel, 1987), 173–83.

174 *Those who saw the same strategic advantage:* A. M. T. Moore and G. C. Hillman, "The Pleistocene to Holocene Transition and Human Economy," 489.

175 *Domestication is the human creation of a new form of plant or animal:* B. D. Smith, *The Emergence of Agriculture* (New York: Scientific American Library, 1995), p. 18.

175 *Unintentional spillage of these larger seeds:* Ibid., 21.

175 *Repeated cycles of cutting, spilling, and growing:* D. Zohary and M. Hopf, *Domestication of Plants in the Old World: The Origin and Spread of Cultivated Plants in West Asia, Europe, and Africa* (Oxford: Oxford University Press, 2nd ed., 1993).

175 *The inadvertently disturbed soil settings close to human settlements:* B. D. Smith, *The Emergence of Agriculture,* p. 21.

176 *on the hilly flanks:* R. J. Braidwood, "The Agricultural Revolution," *Scientific American* 203 (1960): 130–41; H. E. Wright, Jr., "The Environmental Setting for Plant Domestication in the Near East," *Science* 194 (1968): 385–89; D. Zohary and M. Hopf, *Domestication of Plants in the Old World.*

176 *This was an opportunity:* W. B. F. Ryan et al., "Evidence of an Abrupt Submergence of the Black Sea Shelf During the Holocene: Implications for Climate and Diaspora," eds. N. Roberts, M. Karabıyıkoğlu, C. Kuzucuoğlu, *The Late Quaternary in the Eastern Mediterranean—Programme and Abstracts,* Ankara, Turkey (1997).

177 *Lake Akşehir in the west was five times its present size:* N. Kazanci et al., "Paleoclimatic Significance of the Late Pleistocene Deposits of Aksehir

Lake, West-Central Anatolia." Paper presented at The Late Quaternary in the Eastern Mediterranean, Ankara, Turkey, 1997.

177 *the transition from dry to moist:* G. Lemcke, M. Sturm, and L. Wick, "Lake Van (Turkey)—an Ideal Site to Reconstruct Frequent Climatic Fluctuations During the Holocene." Paper presented at The Late Quaternary in the Eastern Mediterranean, Ankara, Turkey, 1997.

177 *Ofer Bar-Yosef from Harvard University gave the keynote paper:* O. Bar-Yosef, "Human Prehistory and Environmental Change in the Eastern Mediterranean (20 ka through 7 ka BP)." Paper presented at The Late Quaternary in the Eastern Mediterranean—Programme and Abstracts, Ankara, Turkey, 1997.

177 *linked the impact of the Younger Dryas:* O. Bar-Yosef and A. Belfer-Cohen, "The Origins of Sedentism and Farming Communities in the Levant," *Journal of World Prehistory* 3, no. 4 (1989): 447–98; O. Bar-Yosef and A. Belfer-Cohen, "From Sedentary Hunter-Gatherers to Territorial Farmers in the Levant," in *Between Bands and States,* ed. S. A. Gregg (Southern Illinois University, Center for Archaeological Investigations, 1991), 181–202; O. Bar-Yosef and A. Belfer-Cohen, "From Foraging to Farming in the Mediterranean Levant," in *Transitions to Agriculture in Prehistory,* ed. A. B. Gebauer and T. D. Trice (Madison, Wis.: Prehistory Press, 1992), 21–48.

177 *The road map came in the very next lecture:* A. N. Goring-Morris, "Late Quaternary settlement patterns and climatic change in the eastern Mediterranean," N. Roberts, M. Karabıyıkoğlu, C. Kuzucuoğlu, eds., *The Late Quaternary in the Eastern Mediterranean—Programme and Abstracts,* Ankara, Turkey (1997).

178 *He presented a strong correlation:* A. N. Goring-Morris, *At the Edge: Terminal Pleistocene Hunter-Gatherers in the Negev and Sinai* (Oxford: BAR International Series, 1987).

178 *These migrations had been tracked in the Negev highlands:* D. O. Henry, "Prehistoric Cultural Ecology in Southern Jordan," *Science* 265, no. 15 (July 1994): 336–41.

178 *In the early 1950s she had gone to Jordan:* K. M. Kenyon, *Digging Up Jericho* (New York: Frederick A. Praeger, 1957).

179 *Among the intricate stone blades:* K. M. Kenyon, *Archaeology in the Holy Land* (New York: Frederick A. Praeger, 2nd ed., 1965), 41.

179 *Years later this obsidian would be subject to chemical analysis:* C. Renfrew, J. E. Dixon, and J. R. Cann, "Further Analysis of Near Eastern Obsidians," *Proceedings of the Prehistoric Society* 34 (1968): 319–31; J.

Keller and C. Seifried, "The Present Status of Obsidian Source Characterization in Anatolia and the Near East," *PACT* 25 (1990): 57–87.

179 *Neolithic village recently excavated:* U. Esin, "Salvage Excavations at the Pre-Pottery Site of Asikli Höyük in Central Anatolia," *Anatolia* 17 (1991): 123–64; U. Esin, "The Aceramic Site of Asikli and Its Ecology." Paper presented at The Late Quaternary in the Eastern Mediterranean, Ankara, Turkey, 1997.

180 *These foundations date back:* U. Esin, "Asikli, Ten Thousand Years Ago: A Habitation Model from Central Anatolia," in *Housing and Settlement in Anatolia—a Historical Perspective* (Istanbul: Tarih Vakfi, 1996), 31–42.

180 *the famous landscape painting from the north and east walls of shrine VII.14 at Çatal Hüyük:* J. Mellaart, *Çatal Hüyük—a Neolithic Town* (New York: McGraw Hill, 1967): figure 59, p. 133.

183 *Çatal Hüyük was reopened:* I. Hodder, ed., *On the Surface: Çatalhöyük 1993–95,* vol. 1, Çatalhöyük Project; BIAA Monograph 22, *McDonald Institute Monographs* (Ankara: British Institute of Archaeology, 1997); O. C. Shane III and M. Küçük, "The World's First City," *Archaeology* (March/April 1998): 43–47.

183 *beheaded corpse on a rack outside the town:* J. Mellaart, *Çatal Hüyük —a Neolithic Town,* figures 45–49, pp. 90–92.

184 *Scientific drilling of the ice cap of Greenland:* W. Dansgaard, S. J. Johnsen, and J. Møller, "One Thousand Centuries of Climatic Record from Camp Century on the Greenland Ice Sheet," *Science* 166, no. 17 (October 1969): 377–81; W. Dansgaard et al., "A New Greenland Deep Ice Core," *Science* 218 (1982): 1273–77; W. Dansgaard et al., "Evidence for General Instability of Past Climate from a 250-kyr Ice-Core Record," *Nature* 364, no. 15 (July 1993): 218–20.

184 *At a depth of 250 feet:* W. Dansgaard, "The Isotopic Composition of Natural Waters, with Special Reference to the Greenland Ice Cap," *Medd. om Gronland,* Bd. 165, pp. 1–120.

184 *a European consortium:* S. J. Johnsen et al., "Irregular Glacial Interstadials Recorded in a New Greenland Ice Core," *Nature* 359 (1992): 311–13.

184 *An American team joined them in 1993:* K. C. Taylor et al., "Electrical Conductivity Measurements from the GISP2 and GRIP Greenland Ice Cores," *Nature* 366, no. 9 (December 1993): 549–52; K. C. Taylor et al., "The 'Flickering Switch' of Late Pleistocene Climate Change," *Nature* 361, no. 4 (February 1993): 432–36.

184 *separate tiny bubbles of fossil air from the ice and measure its methane content:* J. Chappellaz et al., "Synchronous Changes in Atmospheric CH₄ and Greenland Climate between 40 and 8 kyr BP," *Nature* 366, no. 2 (December 1993): 443–45; T. Blinier et al., "Variations in Atmospheric Methane Concentration During the Holocene Epoch," *Nature* 374 (March 1995): 46–49.

184 *started in 6200 B.C.:* R. B. Alley et al., "Holocene Climate Instability: A Large Event 8200 Years Ago," *Geology* 25 (1997): 483–89.

185 *Even Lake Victoria in tropical Africa:* J. C. Stager and P. A. Mayewski, "Abrupt Early to Mid-Holocene Climatic Transition Registered at the Equator and the Poles," *Science* 276 (June 1997): 1834–36.

185 *Rossignol-Strick had recently put together a synthesis:* M. Rossignol-Strick, "Sea-Land Correlation of Pollen Records in the Eastern Mediterranean for the Glacial-Interglacial Transition: Biostratigraphy vs Radiometric Time-Scale," *Quaternal Science Reviews* 14 (1995): 893–915.

185 *compiled by Bulgarian researchers:* V. Filipova, E. D. Bozilova, and P. S. Dimitrov, "Palynological and Stratigraphic Data from the Southern Part of the Bulgarian Black Sea Shelf," *Bulgarian Academy of Sciences Oceanology* (1983): 24–32; J. R. Atanassova and E. D. Bozilova, "Palynological Investigation of Marine Sediments from the Western Sector of the Black Sea," *Proceedings of the Institute of Oceanology,* Varna 1 (1992): 97–103.

185 *a personal letter from one of them:* dated September 3, 1997, addressed to Dr. Rossignol-Strick, and signed by Ellissaveta Bozilova.

185 *goosefoot and wormwood: Chenopodaceae* and *Artemisia,* respectively.

185 *this was when the oaks took hold elsewhere in Europe:* E. Bozilova, "Holocene Chronostratigraphy in Bulgaria," *Striae* 16 (1982): 88–90.

185 *hazel, elm, beech, alder, and birch: Corylus, Ulmus, Fagus, Alnus,* and *Betula.*

186 *"The archaeological evidence points to this conclusion":* A. M. T. Moore, "The Late Neolithic in Palestine," *Levant* 5 (1973): 41.

186 *Plant geneticists announced a match of the DNA:* M. Heun et al., "Site of Einkorn Wheat Domestication Identified by DNA Fingerprinting," *Science* 278, (November 1997): 1312–14.

187 *The fingerprint caught all the experts:* M. K. Jones, R. G. Allaby, and T. A. Brown, "Wheat Domestication (Letters to the Editor)," *Science* 279 (January 1998): 302–03. F. Hole, "Wheat Domestication (Letters to the Editor)," *Science* 279 (January 1998): 303.

187 *malaria and arthritis:* J. L. Angel, "Osteoarthritis in Prehistoric Turkey

and Medieval Byzantium," *Henry Ford Hospital Medical Journal* 27 (1979): 38–43; J. L. Angel, "Early Neolithic Skeletons from Çatal Hüyük: Demography and Pathology," *Anatolian Studies* 21 (1971): 77–98; T. Molleson, "The Eloquent Bones of Abu Hureyra," *Scientific American* (August 1994): 70–75.

17. THE DIASPORA

Page

188 *The strata, peppered with bones, stone tools, and broken pottery:* M. Vasic, *Praistoriska Vinca*, 4 vols. (Belgrade: 1932–36); J. Chapman, *The Vinca Culture of South-East Europe* (Oxford: BAR International Series, 1981).

190 *"as a center of Aegean civilization":* *Illustrated London News* (1930) cited in M. Gimbutas, *The Goddesses and Gods of Old Europe* (London: Thames and Hudson, 1982), 23.

190 *Another group of farmers called Linearbandkeramik:* T. D. Price, A. B. Gebauer, and L. H. Keeley, "The Spread of Farming into Europe North of the Alps," in *Last Hunters—First Farmers: New Perspectives on the Prehistoric Transition to Agriculture,* ed. T. D. Price and A. B. Gebauer (Santa Fe, NM: School of American Research, 1995), 97, 100; P. Bogucki, "The Spread of Early Farming in Europe," *American Scientist* 84 (May–June 1996): 247; G. Barker, "Prehistoric Farming in Europe," in *New Studies in Archaeology,* ed. C. Renfrew and J. Sabloff (Cambridge: Cambridge University Press, 1985), 139–47.

190 *unresolvable by the radiocarbon dating methods:* T. D. Price, A. B. Gebauer, and L. H. Keeley, "The Spread of Farming into Europe North of the Alps," p. 98.

190 *these huge timber-framed houses, up to 150 feet in length:* M. Gimbutas, *The Civilization of the Goddess* (New York: HarperCollins, 1991): 41, figures 2.32, 2.33; A. Hampel, *Die Hausentwicklung im Mittelneolithikum Zentraleuropas* (Bonn: Habelt, 1989).

190 *These dwellings were the largest freestanding buildings:* P. Bogucki, "The Largest Buildings in the World 7,000 Years Ago," *Archaeology* 48, no. 6 (1995): 57–59; S. Milisauskas, *Early Neolithic Settlement and Society at Olszanica* (Ann Arbor: Museum of Anthropology, University of Michigan, 1986); P. Bogucki, "The Spread of Early Farming in Europe," 249, figure 8.

190 *The absence of pigment:* M. Gimbutas, *The Civilization of the Goddess,* p. 37.

191 *A very striking feature of the LBK is the homogeneity:* T. D. Price, A. B.

Gebauer, and L. H. Keeley, "The Spread of Farming into Europe North of the Alps," p. 97; S. A. Greg, *Foragers and Farmers—Population Interaction and Agricultural Expansion in Prehistoric Europe* (Chicago: University of Chicago Press, 1988), 3.

191 *Their explosive movement from east to west up the Dniester and Vistula rivers:* Greg, Ibid., 98; S. Milisauskas, *Early Neolithic Settlement and Society at Olszanica.*

191 *Like the Vinča, the Linearbandkeramik never put down permanent roots near a sea coast:* M. Gimbutas, *The Civilization of the Goddess,* 35, figure 2.28.

191 *the Danilo-Hvar settled on old abandoned sites:* Ibid., 53, 55.

191 *They crafted a now-famous pot decorated with a sailing ship:* Ibid., 55, figure 3.3.

192 *One dubbed "The Thinker":* Ibid., 249, figure 7.42. This figure and its companion are on display at the Museum of National History and Archaeology, Constanta, Romania.

192 *All have been described as outsiders:* D. Berciu, *Cultura Hamangia* (Bucharest: Academia Populare Romine, 1996).

192 *"has led many scholars in the past to assume":* M. Gimbutas, *The Civilization of the Goddess,* 52.

193 *In the mound of Hacılar in western Anatolia:* J. Mellaart, *Çatal Hüyük —a Neolithic Town* (New York: McGraw Hill, 1967), p. 26; J. G. Macqueen, *The Hittites,* rev. ed. (London: Thames & Hudson, 1986), 16–17.

195 *At Hacılar, Mellaart had uncovered five building horizons:* J. Mellaart, *Çatal Hüyük—a Neolithic Town,* 25.

195 *"Considerable changes in painted pottery":* Ibid., 26.

195 *The Halaf were farmers and herders who made beautiful pottery:* G. Roux, *Ancient Iraq,* 2nd ed. (London: Penguin Books, 1980), 67–69.

195 *their buildings were dome shaped:* C. K. Maisels, "The Near East: Archaeology in the 'Cradle of Civilization,' " ed. A. Wheatcroft, *The Experience of Archaeology* (London: Routledge, 1993), 124–26.

196 *A new flint industry was introduced:* F. Wendorf, R. Said, and R. Schild, "Egyptian Prehistory: Some New Concepts," *Science* 169, no. 3951 (1970): 1168; F. Wendorf et al., "Prehistory of the Egyptian Sahara," *Science* 193 (1976): 103–14.

196 *Domesticated cereals and animals with direct genetic affinity:* M. A. Hoffman, *Egypt Before the Pharaohs* (New York: Barnes & Noble, 1979), 102, 181.

197 *In recent years archaeologists have reported:* P. Glumac and D. An-

thony, "The Caucasus," in *Chronologies in Old World Archaeology,* ed. R. Ehrich (Chicago: University of Chicago Press, 1991), 200.

197 *some were dome shaped:* A Dzhavakhishvili, "Construction and Architecture of Southern Caucasus Dwellings, fifth to third millennium B.C." (Tbilisi: 1973); J. Mellaart, *The Neolithic of the Near East* (London: Thames and Hudson, 1975).

197 *dug ditches for simple irrigation:* R. Munchaev, "The Kavkaza Neolithic" in *USSR Neolithic,* ed. V. Masson and N. Merpert (Moscow: Nauka, 1982).

198 *close ties with Turkish, Finnish, and Hungarian:* noted by Jules Oppert as described in S. N. Kramer, *The Sumerians* (Chicago: University of Chicago Press, 1963), 21, and in J. Bottéro, *Initiation à l'Orient ancien, l'histoire* (Paris: Editions du Seuil, 1992), 39.

198 *"we do not know anything of their earlier ties":* J. Bottéro, *Mesopotamia: Writing, Reasoning, and the Gods* (Chicago: University of Chicago Press, 1987), 68.

198 *the seven sages appeared from the sea:* S. Dalley, *Myths from Mesopotamia* (Oxford: Oxford University Press, 1971), 327.

199 *Standing on the summit . . . one can distinguish:* C. L. Woolley, *Ur of the Chaldees* (London: Harmondsworth, 1938), 13.

199 *One of the foundation dwellings was over forty feet in length:* R. J. Matthews and T. J. Wilkinson, "Excavations in Iraq, 1989–1990," *Iraq* 53 (1991): 169–82.

199 *Many of the Sumerian cities, such as Eridu:* C. K. Maisels, *The Emergence of Civilization* (London: Routledge, 1990), 135.

200 *The southern Mesopotamians . . . had big, long and narrow heads:* Sir Arthur Keith cited in a footnote of C. L. Wooley, *The Sumerians* (New York: W. W. Norton, 1965), 6–7.

200 *most of the root words are monosyllabic:* S. N. Kramer, *The Sumerians* (Chicago: University of Chicago Press, 1963), 41; C. B. F. Walker, *Cuneiform* (Berkeley: University of California Press, 1987), 11–12.

200 *those having to do with agriculture and crafts are polysyllabic:* S. N. Kramer, *The Sumerians,* 41.

200 *Linguists have discovered that languages:* A. C. Renfrew, *Archaeology and Language: The Puzzle of Indo-European Origins* (Cambridge: Cambridge University Press, 1988), 99–119; J. P. Mallory, *In Search of the Indo-Europeans: Language, Archaeology, and Myth* (Thames and Hudson, 1989), 9–23; T. V. Gamkrelidze and V. V. Ivanov, "The Early History of Indo-European Languages," *Scientific American* (March 1990): 110–11.

18. FAMILY TREES

Page

202 *"It was a beautifully preserved roll of paper"*: M. A. Stein, *Ruins of Desert Cathay: Personal Narrative of Explorations in Central Asia and Westernmost China* (Delhi: B. R. Publishing Company, 1912, 1985), 29.

203 *A naturally mummified woman:* Results of Victor H. Mair of the University of Pennsylvania Museum and Dolkum Kamberi of Tarim, China, as reported in J. N. Wilford, "Mummies, Textiles Offer Evidence of Europeans in Far East," *New York Times,* Tuesday, May 7, 1996, C8.

204 *"virtually identical stylistically and technically"*: Quotation of Irene Good of the University of Pennsylvania Museum reported in J. N. Wilford, "Mummies, Textiles Offer Evidence of Europeans in Far East," *New York Times,* Tuesday, May 7, 1996, C8.

204 *"What makes individuals and populations biologically different"*: L. L. Cavalli-Sforza and F. Cavalli-Sforza, *The Great Human Diasporas* (New York: Addison-Wesley, 1995), 92.

205 *"Our findings fully confirmed our expectations"*: Ibid., 101.

205 *"It may seem strange"*: Ibid., 101.

205 *Most groups do not stay isolated:* Ibid., 103–04.

207 *"if enough data on a number of different genes are gathered"*: Ibid., 111–12.

207 *The picture seemed to show a wave of advance:* A. J. Ammerman and L. L. Cavalli-Sforza, "Measuring the Rate of Spread of Early Farming in Europe," *Man* 6 (1971): 674–78; A. J. Ammerman and L. L. Cavalli-Sforza, *The Neolithic Transition and the Genetics of Populations in Europe* (Princeton: Princeton University Press, 1984).

207 *their length would point to the time since fissioning:* R. R. Sokal, N. L. Oden, and C. Wilson, "Genetic Evidence for the Spread of Agriculture in Europe by Demic Diffusion," *Nature* 351 (1991): 143–45.

207 *The project to construct a genetic tree:* L. L. Cavalli-Sforza et al., "Reconstruction of Human Evolution: Bringing Together Genetic, Archaeological, and Linguistic Data," *Proceedings of the National Academy of Sciences* 85 (August 1988): 6003; L. L. Cavalli-Sforza, E. Minch, and J. L. Mountain, "Coevolution of Genes and Languages Revisited," *Proceedings of the National Academy of Sciences* 89 (June 1992): 5621.

208 *"if the tree of genetic evolution were known"*: C. Darwin, *On the Origin of the Species by Means of Natural Selection, or the Preservation of Favored Races in the Struggle for Life* (London: Murray, 1859).

208 *"some common source, which perhaps no longer exists"*: Sir W. Jones,

"Third Anniversary Discourse: On the Hindus" (1786), reprinted in *A Reader in the Nineteenth-Century Historical Indo-European Linguistics*, ed. W. P. Lehmann (Bloomington: Indiana University Press, 1967), 15.

208 *"the strong affinity, both in the roots of verbs and in the forms of grammar"*: Ibid., 15.

209 *"Linguists seek correspondence in grammar, syntax, vocabulary, and vocalization"*: T. V. Gamkrelidze and V. V. Ivanov, "The Early History of Indo-European Languages," *Scientific American* (March 1990): 110.

210 *the proto-Indo-European had words for bear*: T. Gamkrelidze and V. V. Ivanov, "The Ancient Near East and the Indo-European Question: Temporal and Territorial Characteristics of Proto-Indo-European Based on Linguistic and Historico-Cultural Data," *Journal of Indo-European Studies* 13 (1985): 9–21; J. P. Mallory, *In Search of the Indo-Europeans: Language, Archaeology, and Myth* (London: Thames and Hudson, 1989): 115–16.

210 *The homeland, so defined by the nature of its landscape*: Ibid., 144, figure 80; A. C. Renfrew, "The Origins of Indo-European Languages," *Scientific American*, (October 1989): 114; T. V. Gamkrelidze and V. V. Ivanov, "The Early History of Indo-European Languages," 112.

210 *The experts who attached the Kurgan warriors*: M. Gimbutas, "The First Wave of European Pastoralists into Copper Age Europe," *Journal of Indo-European Studies* 5, no. 4 (1977): 277–399.

210 *And those who envisioned the spread of language moving in tandem with the wave of agriculture*: A. J. Ammerman and L. L. Cavalli-Sforza, "Measuring the Rate of Spread of Early Farming in Europe," *Man* 6 (1971): 674–78; A. J. Ammerman and L. L. Cavalli-Sforza, *The Neolithic Transition and the Genetics of Populations in Europe* (Princeton: Princeton University Press, 1984); L. L. Cavalli-Sforza and F. Cavalli-Sforza, *The Great Human Diasporas*, figure 6.5; A. C. Renfrew, *Archaeology and Language: The Puzzle of Indo-European Origins* (Cambridge: Cambridge University Press, 1988).

211 *pronunciation changes that were contrary to the classical rules*: T. V. Gamkrelidze and V. V. Ivanov, "The Early History of Indo-European Languages," 110–13.

211 *"strayed light-years away from whatever consensus"*: J. P. Mallory, *In Search of the Indo-Europeans: Language, Archaeology, and Myth*, 7.

211 *a team of three Americans*: The research of Donald Ringle, Ann Taylor, and Tandy Warrow at the University of Pennsylvania was presented at the National Academy of Science Symposium on the Frontiers of Sci-

ence in Irvine, California, in November 1995 and reported by G. Johnson, "New Family Tree Is Constructed for Indo-European," *New York Times,* Jan. 2, 1996, B15.

212 *"It might be fair to say that I was biased against it":* Ibid.

212 *"So you can imagine how startled I was":* Ibid.

212 *Semitic, Kartvelian, Sumerian, and even Egyptian:* T. V. Gamkrelidze and V. V. Ivanov, "The Early History of Indo-European Languages," 115.

212 *The residue of retsina wine:* P. E. McGovern et al., "Neolithic Resinated Wine," *Nature* 381, no. 6 (June 1996): 480–81.

213 *"a dead forest of sun-bleached, wind-scoured tree stumps":* K. Wimmel, *Alluring Target* (Washington, DC: Trackless Sands Press, 1996), 22.

214 *was one of the most unexpected discoveries that I made:* Ibid., 23.

215 *"the highest coastline was discontinuous and not very fresh":* O. Erol et al., "Was the Tarim Basin Occupied Entirely by a Giant Late Pleistocene Lake?" *Abstracts with Programs of the 1996 Annual Meeting of the Geological Society of America* 28, no. 7 (1996): 497.

19. Guslar's Song

Page

217 *Ever since his arrival in Novi Pazar:* M. Knight, "Homer in Bosnia," *The Sciences* (March/April 1993): 10.

217 *"still alive on the lips of men":* J. M. Foley, "Tradition-dependent and -independent Features in Oral Tradition: A Comparative View of the Formula," in *Oral Tradition Literature,* ed. J. M. Foley (Columbus: Slavica Publishers: 1980): 271.

220 *"a group of words regularly employed":* Milman Parry quoted in M. Knight, "Homer in Bosnia," 11.

221 *Parry had found no less than twenty-nine formulas:* Ibid., 11.

221 *"I say! In this way I have heard it":* Ibid., 12.

222 *"The patterns must be suprahistorical":* Albert Lord quoted in J. M. Foley, *The Theory of Oral Composition: History and Methodology,* 47.

222 *"Fact is present in the epic":* Ibid.

222 *"a helmet fashioned of leather":* R. Lattimore, *The Iliad of Homer* (Chicago: University of Chicago Press, 1951), 225 (book 10, verses 257–65).

222 *This same boar's tusk helmet:* O. Negbi, "The 'Miniature Fresco' fom Thera and the Emergence of Mycenaean Art," in *Thera and the Aegean World,* ed. C. Doumas (London: Thera and the Aegean World, 1978), p. 646.

222 *Lord considered the Gilgamesh epic:* H. Mason, *Gilgamesh* (New York: Signet, 1972), 108.

224 *"stitch-up job":* H. McCall, *Mesopotamian Myths* (Austin: University of
     Texas Press, 1990), 35.
224 *The creation story that George Smith:* W. G. Lambert and A. R. Millard,
     Atra-hasis: The Babylonian Story of the Flood (Oxford: Oxford Univer-
     sity Press, 1969), 7.
224 *One Sumerian poem:* B. Alster, *Dumuzi's Dream: Aspects of Oral Poetry
     in a Sumerian Myth,* vol. 1, *Mesopotamia: Copenhagen Studies in Assy-
     riology* (Copenhagen: Akademisk Forlag, 1972), 13.
224 *The premise for considering that the Sumerian poems:* W. G. Lambert
     and A. R. Millard, *Atra-hasis: The Babylonian Story of the Flood,* 8.
224 *Within a single version:* B. Alster, *Dumuzi's Dream: Aspects of Oral
     Poetry in a Sumerian Myth,* vol. 1, *Mesopotamia,* 16–17.
224 *He has sprouted, he has burgeoned:* Translated by S. N. Kramer and
     cited in J. B. Pritchard, *The Ancient Near East,* 2 vols., (Princeton:
     Princeton University Press, 1975), vol. 2, p. 202.
225 *According to the Sumerian king-list:* S. N. Kramer, *The Sumerians* (Chi-
     cago: University of Chicago Press, 1963), p. 328, Appendix E.
225 *The adaptation of his recorded exploits:* W. G. Lambert and N. K. Sand-
     ers, *The Epic of Gilgamesh* (Middlesex, England: Penguin Books, 1960),
     13. There are five known Sumerian stories celebrating events of his life
     *(Gilgamesh and the Bull of Heaven; The Death of Gilgamesh; Gil-
     gamesh and Agga of Kish; Gilgamesh and the Land of the Living; Gil-
     gamesh, Enkidu, and the Nether World),* which are partly translated.
     S. N. Kramer, *The Sumerians,* 185–205.
225 *But even then continued oral performance:* "The Historicity of Gil-
     gamesh," in *Gilgamesh and his legend,* ed. P. Garelli (Paris: 1960), 50.

2 1 .   O t h e r   M y t h s
Page
239 *ossiferous caverns:* W. Buckland, *Reliquiae Diluvianae, or, Observa-
     tions on the Organic Remains Contained in Caves, Fissures, and Dilu-
     vial Gravel, and on Other Geological Phenomena Attesting to the Action
     of an Universal Deluge* (London: John Murray, 1823).
239 *"The details given in the inscriptions":* G. Smith, *The Chaldean Account
     of Genesis* (New York: Scribner, Armstrong & Co., 1876), 307.
240 *"a little hidden door":* C. G. Jung, "The Meaning of Psychology for
     Modern Man in Civilization in Transition," in *The Collected Works of
     C. G. Jung* (Princeton: Princeton University Press, 1970), pp. 304–05,
     as referenced in J. Campbell, *The Mythic Image* (Princeton: Princeton
     University Press, 1990), p. 7.

240 *the material of myth is the material of life:* Campbell, *Transformations of Myth Through Time* (New York: HarperCollins, 1990), p. 1.

240 *trees of the species called* Survan: Ibid., 208.

240 *"in one inscription":* Ibid., 208.

240 *the location of the Humbaba's dwelling:* R. Temple, *He Who Saw Everything* (London: Rider, 1991), 35.

240 *due to a reference to the sun god, Shamash:* H. McCall, *Mesopotamian Myths* (Austin, TX: University of Texas Press, 1990), 41.

240 *the Cedar Mountain is explicitly located in the northwest, in or near Lebanon:* N. K. Sanders, *The Epic of Gilgamesh* (Middlesex, England: Penguin Books, 1960), p. 55; J. H. Tigay, *The Evolution of the Gilgamesh Epic* (Philadelphia: University of Pennsylvania Press, 1982), 78.

240 *"There Gilgamesh dug a well before the setting sun":* N. K. Sanders, *The Epic of Gilgamesh,* 77.

240 *"dense is the darkness, and light there is none":* M. G. Kovacs, *The Epic of Gilgamesh* (Stanford, CA: Stanford University Press, 1989), 76, tablet ix, col. iii.

241 *as if following the way of the sun's night journey:* D. Ferry, *Gilgamesh* (New York: Noonday Press, 1992), 52.

242 *"whose path never sees the light of the sun":* M. L. Settle, *Turkish Reflections* (New York: Touchstone, 1991), 187.

242 *Xenophon led his mercenary Greek army of ten thousand:* Xenophon, *Anabasis,* IV.3, translated in G. C. Cawkwell, *Xenophon: The Persian Expedition* (London: Penguin Books, 1972).

242 *"Gilgamesh, there has never been a crossing":* J. Gardner and J. Maier, *Gilgamesh. Translated from the Sîn-leqi-unninni Version* (New York: Vintage Books, 1985), 212, tablet x, col. ii.

242 *"there slide the other waters, the waters of death":* D. Ferry, *Gilgamesh,* 57.

242 *it has risen by more than eighty feet:* J. W. Murray, "Unexpected Changes in the Oxic/Anoxic Interface in the Black Sea," *Nature* 337 (1989): 411; L. A. Codispoti et al., "Chemical Variability in the Black Sea: Implications of Data Obtained with a Continuous Vertical Profiling System That Penetrated the Oxic-Anoxic Interface," *Deep-Sea Research* 38 (1991): S694.

242 *"anoxygenic photosynthesis":* S. G. Wakeham and J. A. Beier, "Fatty Acid and Sterol Biomarkers as Indicators of Particulate Matter Source and Alteration Processes in the Black Sea," *Deep-Sea Research* 38, suppl. 2 (1991): S959.

243 *"that I might not touch the waters of death":* J. H. Tigay, *The Evolution*

of the Gilgamesh Epic (Philadelphia: University of Pennsylvania Press, 1982), 114.

244 *"senders . . . without which there is no crossing death's waters"*: J. Gardner and J. Maier, *Gilgamesh. Translated from the Sîn-leqi-unninni Version*, 213, tablet x, col. ii.

244 *"talismans"*: D. Ferry, *Gilgamesh*, 60.

244 *images:* A. Heidel, *The Gilgamesh Epic and Old Testament Parallels* (1946), 76 (tablet x, col. iii).

244 *"lodestones"*: R. Temple, *He Who Saw Everything* (London: Rider, 1991), 109, 116, footnote 5.

244 *pulled out their retaining ropes:* M. G. Kovacs, *The Epic of Gilgamesh*, 89.

244 *the protective guardian of the things of stone:* J. Gardner and J. Maier, *Gilgamesh. Translated from the Sîn-leqi-unninni Version*, 215.

244 *connected with the Egyptian word* Urnes: R. Temple, *He Who Saw Everything*, 117, footnote 6.

244 *"there is no word of advice"*: J. Gardner and J. Maier, *Gilgamesh. Translated from the Sîn-leqi-unninni Version*, 222 (tablet x, col. v).

244 *"As for death, its time is hidden"*: Ibid., 224 (tablet x, col. vi).

245 *"If you get your hands on that plant"*: Ibid., 249 (tablet xi, col. vi).

245 *a torpid and gloomy existence forever:* J. Bottéro, *Mesopotamia: Writing, Reasoning, and the Gods* (Chicago: University of Chicago Press, 1987), 230.

245 *There is an older Sumerian version of the flood:* M. Civil, "The Sumerian Flood Story," in *Atra-hasis: The Babylonian Story of the Flood*, ed. W. G. Lambert and A. R. Millard (Oxford: Clarendon Press, 1969), 138–45; translation by S. N. Kramer in J.B. Pritchard, *The Ancient Near East*, 2 vols. (Princeton: Princeton University Press, 1973) vol. 1, 28–30.

245 *The flood had swept over the land:* The Deluge, lines 205–206, translated by S. N. Kramer in J. B. Pritchard, *The Ancient Near East*, 2 vols., vol. 1 (Princeton: Princeton University Press, 1973), 30.

246 *"brought his rays into the giant boat"*: Ibid., line 208.

246 *"life like that of a god"*: Ibid., line 256.

246 *written no earlier than the ninth century B.C.:* B. B. Trawick, *The Bible as Literature*, 2nd ed. (New York: Barnes & Noble, 1970).

247 *"For my part, I am going to bring a flood"*: Ibid., 6:17

247 *"smelled a sweet savour"*: Ibid., 8:21.

247 *in the excavations of Shuruppak:* H. Peake, *The Flood* (London: Kegan Paul, Trench, Trubner & Co., Ltd., 1930), p. 97.

249 *"Dismantle your house, build a boat"*: Gilgamesh, tablet xi, col. i, in S.

Dalley, *Myths from Mesopotamia* (Oxford: Oxford University Press, 1971), 110.

249 *"I loaded her with everything there":* Ibid., tablet xi, col. ii, p. 111.

250 *"The roaring waters of the Deep arose":* R. Graves and R. Patai, *Hebrew Myths* (New York: Doubleday, 1989), p. 29.

250 *"making a decree which Tehom could never break":* Ibid., p. 29.

250 *"Leviathan's monstrous tusks spread terror":* Ibid., p. 47.

250 *The first and original inhabitants used an ancient language:* Diodorus Siculus, *Bibliotheca Historica,* trans. C. H. Oldfather.

251 *"sprung from the rock":* Diodorus Siculus, *Bibliotheca Historica,* trans. C. H. Oldfather.

251 *"like the bellowing of a bull, like a wild ass screaming":* R. Graves and R. Patai, *Hebrew Myths* (New York: Doubleday, 1989), p. 112.

EPILOGUE: A TELLING OF *ATRAHASIS*

Page

253 Atrahasis: Perhaps the oldest preserved account of the flood.

253 *twelfth year of the reign of King Ammi-saduqa of Assyria:* is placed in the First Dynasty of Babylon, circa 1634 B.C. (Old Babylonian period) using the Mesopotamian chronology of J. Oates, *Babylon,* rev. ed. (New York: Thames & Hudson Inc., 1986), p. 200. The *Atrahasis* myth precedes the epic of *Gilgamesh* by at least three centuries.

253 *the land of Hana:* encompasses the region around Terqa in eastern Syria where the Soviets built a dam across the Euphrates in 1974, impounding a large lake and flooding the Neolithic village of Abu Hureyra.

253 *the dry steppe to Yamhad:* refers to ancient Aleppo.

253 *the coast of the azure sea littered with its coveted shells of cowry and conch:* implies the Levant margin of the Mediterranean Sea.

253 *three leagues wide:* In the Akkadian measurement system this is about nine miles or fifteen kilometers.

253 *copper and ivory of Tilmun:* designates the region of Bahrain on the Arabian side of the Persian Gulf.

256 *Six hundred years, less than six hundred:* This and all the other quoted text of *Atrahasis* is from S. Dalley, *Myths from Mesopotamia* (Oxford: Oxford University Press, 1969), 9–35.

Index

Page numbers in *italics* refer to illustrations.

Made in United States
Orlando, FL
20 April 2023